ACE
GENERAL CHEMISTRY
I AND II

(THE EASY GUIDE TO ACE GENERAL CHEMISTRY I AND II)

BY: DR. HOLDEN HEMSWORTH

Copyright © 2015 by Holden Hemsworth

DISCLAIMER

Chemistry, like any field of science, is continuously changing and new information continues to be discovered. The author and publisher have reviewed all information in this book with resources believed to be reliable and accurate and have made every effort to provide information that is up to date and correct at the time of publication. Despite our best efforts we cannot guarantee that the information contained herein is complete or fully accurate due to the possibility of the discovery of contradictory information in the future and any human error on part of the author, publisher, and any other party involved in the production of this work. The author, publisher, and all other parties involved in this work disclaim all responsibility from any errors contained within this work and from any results that arise from the use of this information. Readers are encouraged to check all information in this book with institutional guidelines, other sources, and up to date information.

The information contained in this book is provided for general information purposes only and does not constitute medical, legal or other professional advice on any subject matter. The information author or publisher of this book does not accept any responsibility for any loss which may arise from reliance on information contained within this book or on any associated websites or blogs.

WHY I CREATED THIS STUDY GUIDE

In this book I try to breakdown the content covered in the typical two semester General Chemistry course in college for easy understanding and to point out the most important subject matter that students are likely to encounter. This book is meant to be a supplemental resource to lecture notes and textbooks to boost your learning and go hand in hand with your studying!

I am committed to providing my readers with books that contain concise and accurate information and I am committed to providing them tremendous value for their time and money.

Best regards,

Dr. Holden Hemsworth

TABLE OF CONTENTS

CHAPTER 1 – INTRODUCTION TO CHEMISTRY

What is Chemistry?

Chemistry is the branch of science concerned with the understanding of matter; the substances it is composed of and their properties, as well as the ways in which they interact and change to form new substances.

Matter

Matter is anything that has mass and takes up space. Mass is the amount of matter an object contains; a way of quantifying matter. Matter exists in three physical states.

- Solid – matter with fixed shape and volume (rigid)

- Liquid – matter with a fixed volume but indefinite shape

    o Takes on the shape of the container it is in

- Gas – matter without a fixed shape or volume

    o Conforms to the volume and shape of its container

Physical and Chemical Properties

- Physical property – characteristics that can be measured and observed without changing the chemical makeup of the substance

    o Examples: color, melting point, boiling point, density, etc.

- Physical change – a substance changes its physical appearance but does not change identity

    o Changes in state (e.g., liquid to gas, solid to liquid) are all physical changes

- Chemical property – any property that becomes evident during a chemical reaction

    o Examples: pH, corrosiveness, etc.

- Chemical change (aka chemical reactions) – a substance is transformed into a chemically different substance

Mixtures

Mixtures are combinations of two or more substances in which each substance keeps its chemical identity. Mixtures can be separated into two or more substances.

- Heterogenous mixtures – mixture that is divided into different regions of appearance and properties
    - Results from components not being distributed uniformly
- Homogenous mixtures – mixture that is uniform throughout without any visible separations
    - Solutions are homogenous mixtures
        - Where a solid (the solute) is dissolved in a liquid (the solvent)

Elements and Compounds

Pure substances have definite and consistent composition and are composed of elements or compounds.

- Element - substance that can't be broken down into other substances by chemical means
- Compound – substance formed from two or more chemical elements that are chemically bonded together
- Law of definite proportions
    - Pure compounds always contain exactly the same proportions of elements by mass

Energy

Energy is the capacity to do wok.

- Kinetic energy – energy possessed by an object due to its motion
- Potential energy – energy stored in matter because of its position or location
    - Something suspended in the air has higher potential energy than something sitting on the ground
- Total Energy = potential energy + kinetic energy
- Lower energy states are more stable in nature
- Law of conservation of energy
    - Energy can't be created or destroyed…but it can be transformed
    - Energy is always conserved

Scientific Method

The scientific method is a technique for investigation that is used to answer scientific questions.

- Hypothesis - a proposed explanation made on observations or limited evidence that serves as the starting point for further investigation

- Theory – explanation of general principles of a phenomena that has been repeatedly tested and observed

- Fact – indisputable truth

- Steps in scientific approach

  - Observations, Hypothesis, Experiment, Development of a model or theory, Further experimentation

Measurements

Measured quantities consist of a number and a unit.

- Units are standardized in the form of the International System called SI units

- Units have associated prefixes to make them easier to use and reports

Tera - 10^{12}	Centi – 10^{-2}
Giga – 10^{9}	Milli – 10^{-3}
Mega – 10^{6}	Micro – 10^{-6}
Kilo – 10^{3}	Nano – 10^{-9}
Deci – 10^{-1}	Pico – 10^{-12}

- Conversion factors – a mathematical multiplier used to convert a quantity expressed in one set of units into an equivalent quantity expressed in

  - Example: 1 yard = 3 feet (10 yard = 30 feet)

Scientific Notation

- Scientific notation is a way of handling very large or very small numbers

- Scientific notation for a number only contains significant figures

  - Examples

    - $525,000 = 5.25 \times 10^{5}$

    - $2,301,000,000 = 2.301 \times 10^{9}$

    - $0.000000000670 = 6.70 \times 10^{-10}$

- Consider the following: $0.000023 = 2.3 \times 10^{-5}$
  - The exponent on 10 is the number of places the decimal point must be shifted to give the number in its long form
    - Positive exponent, shift the decimal point to the right
    - Negative exponent, shift the decimal point to the left

Significant Figures

- All non-zero numbers are always significant
  - 1, 2, 3, 4, 5, 6, 7, 8, and 9
- Zeroes in between non-zero numbers are always significant
  - 10001 – 5 sig figs
- A final zero or trailing zeroes in the **decimal portion** are significant
  - 0.00**500** – 3 sig figs
  - 1255.0 – 5 sig figs
- All zeroes to the left of a decimal point and that add value to a number are significant
  - 100.0 – 4 sig figs
  - 0.1 – 1 sig fig
    - In this case, the zero adds no value; it is there to avoid confusion and by convention

Exact Number

- Exact numbers are considered to have an infinite number of significant figures
  - They do not affect accuracy or precision of an expression they are in
  - You do not have to consider the significant figures in exact numbers when doing calculations
- Conversion factors are exact numbers
  - 1 yard = 3 feet (there are exactly 3 feet in a yard)
  - 1 foot = 12 inches (there are exactly 12 inches in a foot)

Multiplication and Division Significant Figures

- First perform all operations and arrive at an answer

- The answer should have the same number of significant figures as the number with the least amount of significant figures used in the calculations

Addition and Subtraction Significant Figures

- First perform all operations and arrive at an answer

- In addition and subtraction you only have to consider the significant figures in the decimal portion

    - The answer should contain no more decimal places than the number with the least amount of digits in the decimal portion

Multiplication/Division Combined with Addition/Subtraction

- Follow order of operations

- If the next operation to be performed is in the same group as the previous operation then don't round the calculation

    - For example when you perform division and then multiplication, you would not round the calculation

- If the next operation to be performed is in the other group from the previous operation then you would round the answer using the rules before moving on to the next operation

    - Example: You perform division and the next operation is subtraction

        - You would first round the result of the division using the significant figure rules for division before you perform subtraction

Accuracy and Precision

- Accuracy – how close a result is to the real value

- Precision – how close repeated measurements are in relation to one another

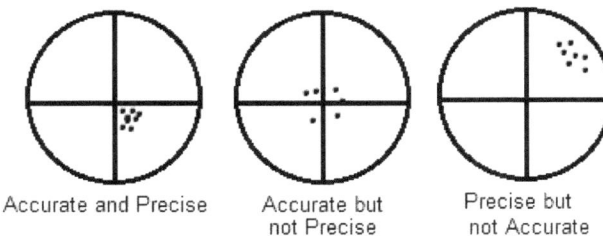

Accurate and Precise   Accurate but not Precise   Precise but not Accurate

Accuracy vs. Precision:

Uncertainty

- Uncertainty – error in a measurement

    o Expressed as a standard deviation

- When making a measurement involving an instrument, the measurement is made with one uncertain digit

    o Example: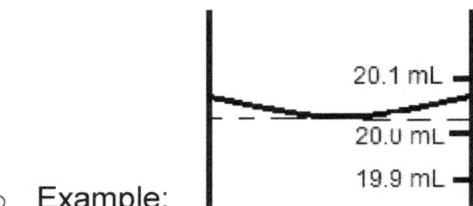

        ▪ You might record the measurement as 20.03

        ▪ The 3 is an uncertain digit because it is estimated and can't be read off exactly from the instrument

Temperature

Temperature is commonly quantified using the three units: kelvin, Celsius, and Fahrenheit.

- Kelvin (K) – "absolute temperature scale"

    o Starts at absolute zero

    o Contains only positive values

- Celsius (°C) – "water based scale"

    o 0°C – freezing point of water

    o 100°C – boiling point of water

    o Most commonly used scale around the world

- Fahrenheit (°F) – "mercury based scale"

    o Commonly used in the US

- Converting Temperatures
    - Formula for Kelvin to Fahrenheit: (9/5)(K - 273) + 32
    - Formula for Kelvin to Celsius: K – 273
    - Formula for Celsius to Kelvin: °C + 273
    - Formula for Celsius to Fahrenheit: °C x (9/5) + 32
    - Formula for Fahrenheit to Kelvin: (5/9)(°F – 32) + 273
    - Formula for Fahrenheit to Celsius: (5/9)(°F - 32)

CHAPTER 2 – COMPONENTS OF MATTER

Components of Matter (Definitions)

- Element - substance that can't be broken down into other substances by chemical means

- Molecule - a combination of two or more atoms

- Compound – substance formed from two or more chemical elements that are chemically bonded together

- Mixture - two or more elements (or compounds) mingling without any chemical bonding

Laws of Matter

- Law of Mass Conservation

    o Total masses of substances involved in a chemical reaction do not change

        ▪ Number of substances and their properties can change

- Law of Definite Proportions:

    o Pure compounds contain exactly the same proportions of elements by mass

- Law of Multiple Proportions

    o If two elements react to form more than one compound, then the ratios of the masses of the second element which combine with a fixed mass of the first element will be in ratios of small whole numbers

Postulates of Dalton's Atomic Theory

- All matter consists of extremely small particles called atoms

- All atoms of an element are identical

    o They are different from atoms of any other element

    o Including in mass and other properties

- Atoms of an element can't be converted into atoms of another element

- Compounds result when atoms of more than one element combine

Periodic Table of Elements

The periodic table is an arrangement of elements in rows and columns based on their atomic number, electron configurations, and chemical properties.

- Period – horizontal row on the table

- Group (Family) – column on the table

hydrogen 1 H 1.0079																	helium 2 He 4.0026
lithium 3 Li 6.941	beryllium 4 Be 9.0122											boron 5 B 10.811	carbon 6 C 12.011	nitrogen 7 N 14.007	oxygen 8 O 15.999	fluorine 9 F 18.998	neon 10 Ne 20.180
sodium 11 Na 22.990	magnesium 12 Mg 24.305											aluminium 13 Al 26.982	silicon 14 Si 28.086	phosphorus 15 P 30.974	sulfur 16 S 32.065	chlorine 17 Cl 35.453	argon 18 Ar 39.948
potassium 19 K 39.098	calcium 20 Ca 40.078	scandium 21 Sc 44.956	titanium 22 Ti 47.867	vanadium 23 V 50.942	chromium 24 Cr 51.996	manganese 25 Mn 54.938	iron 26 Fe 55.845	cobalt 27 Co 58.933	nickel 28 Ni 58.693	copper 29 Cu 63.546	zinc 30 Zn 65.39	gallium 31 Ga 69.723	germanium 32 Ge 72.61	arsenic 33 As 74.922	selenium 34 Se 78.96	bromine 35 Br 79.904	krypton 36 Kr 83.80
rubidium 37 Rb 85.468	strontium 38 Sr 87.62	yttrium 39 Y 88.906	zirconium 40 Zr 91.224	niobium 41 Nb 92.906	molybdenum 42 Mo 95.94	technetium 43 Tc [98]	ruthenium 44 Ru 101.07	rhodium 45 Rh 102.91	palladium 46 Pd 106.42	silver 47 Ag 107.87	cadmium 48 Cd 112.41	indium 49 In 114.82	tin 50 Sn 118.71	antimony 51 Sb 121.76	tellurium 52 Te 127.60	iodine 53 I 126.90	xenon 54 Xe 131.29
caesium 55 Cs 132.91	barium 56 Ba 137.33	57-70 ✳	lutetium 71 Lu 174.97	hafnium 72 Hf 178.49	tantalum 73 Ta 180.95	tungsten 74 W 183.84	rhenium 75 Re 186.21	osmium 76 Os 190.23	iridium 77 Ir 192.22	platinum 78 Pt 195.08	gold 79 Au 196.97	mercury 80 Hg 200.59	thallium 81 Tl 204.38	lead 82 Pb 207.2	bismuth 83 Bi 208.98	polonium 84 Po [209]	astatine 85 At [210]	radon 86 Rn [222]
francium 87 Fr [223]	radium 88 Ra [226]	89-102 ✳✳	lawrencium 103 Lr [262]	rutherfordium 104 Rf [261]	dubnium 105 Db [262]	seaborgium 106 Sg [266]	bohrium 107 Bh [264]	hassium 108 Hs [269]	meitnerium 109 Mt [268]	ununnilium 110 Uun [271]	unununium 111 Uuu [272]	ununbium 112 Uub [277]	ununquadium 114 Uuq [289]				

✳Lanthanide series	lanthanum 57 La 138.91	cerium 58 Ce 140.12	praseodymium 59 Pr 140.91	neodymium 60 Nd 144.24	promethium 61 Pm [145]	samarium 62 Sm 150.36	europium 63 Eu 151.96	gadolinium 64 Gd 157.25	terbium 65 Tb 158.93	dysprosium 66 Dy 162.50	holmium 67 Ho 164.93	erbium 68 Er 167.26	thulium 69 Tm 168.93	ytterbium 70 Yb 173.04
✳✳Actinide series	actinium 89 Ac [227]	thorium 90 Th 232.04	protactinium 91 Pa 231.04	uranium 92 U 238.03	neptunium 93 Np [237]	plutonium 94 Pu [244]	americium 95 Am [243]	curium 96 Cm [247]	berkelium 97 Bk [247]	californium 98 Cf [251]	einsteinium 99 Es [252]	fermium 100 Fm [257]	mendelevium 101 Md [258]	nobelium 102 No [259]

- Elements on the periodic table can be classified as metals, nonmetals, and metalloids

    - Metal – substances that have luster, high heat conductivity, high electrical conductivity, and are solid at room temperature (exception: mercury)

    - Nonmetal – substance without any metal characteristics

    - Metalloid – substance that have both metal and nonmetal characteristics

Atoms

An atom is the smallest unit of matter. Atoms interact to form molecules. Atoms are composed of subatomic particles (electrons, protons, and neutrons).

- Electrons – negatively charged particles

    - Carries a charge of -1.602×10^{-19} Coulombs (C)

- - Charge of atomic and sub-atomic particles are typically described as a multiple of this value

    - So, referred to as -1

  - Mass = $9.10938291 \times 10^{-31}$ kg

- Protons – positively charged particles

  - Carries a charge of $+1.602 \times 10^{-19}$ Coulombs (C)

    - Referred to as a +1 electron charge

  - Mass = $1.67262178 \times 10^{-27}$ kg

- Neutrons – uncharged particles

  - Electrically neutral

  - Mass = $1.674927351 \times 10^{-27}$ kg

- Protons and neutrons are found in the nucleus

  - Nucleus is the central core of an atom

- Electrons orbit the nucleus in an "electron cloud"

- Elemental (atomic) symbol: shorthand representation of atoms of different elements

11	———	Atomic number
Na	———	Element symbol
Sodium	———	Element name
22.990	———	Atomic weight

Example of an Element on the Periodic Table:

- Atomic number - number of protons in an atom of a particular element

  - For a neutral atom, number of electrons = number of protons

  - All atoms of an element have the same atomic number (same number of protons)

- Mass number = the number of protons + the number of neutrons

  - All atoms of an elements don't have the same number of neutrons

- Atomic weight (relative atomic mass) – average mass of atoms of an element
    - Calculated based on the relative abundance of isotopes in that particular element
    - Units: atomic mass units (amu)
- Isotopes – atoms of an element with the same number of protons but with a different number of neutrons
    - Same atomic mass but different mass number

Types of Chemical Formulas

Chemical formulas are a way of expressing information about the proportions of atoms that constitute a compound using: element symbols, numerical subscripts, and other symbols (e.g., parentheses, dashes).

- Empirical formula – smallest whole number ratio of numbers of the atoms in a molecule
- Molecular formula – actual number of atoms in a molecule
- Structural formula – chemical formula showing how atoms are bonded together in a molecule

Covalent and Ionic Bonds

Covalent Bonds

- Two atoms share their valence electrons (electrons in the outer shell of an atom)
- Two Types
    - Non-polar covalent bond – electrons shared equally between atoms
        - Electronegativity of the two atoms is about the same
        - Typically electronegativity difference between the two atoms has to be less than 0.5 for non-polar bonds
        - Electronegativity – an atom's ability to attract and hold on to electrons, represented by a number
    - Polar covalent bonds – electrons shared disproportionately between atoms
        - Electronegativity between the two atoms is different by a greater degree than 0.5 but less than 2.0

Ionic Bonds

- Electrons are transferred, not shared between atoms
- An atom with high electronegativity will take an electron from an atom with low electronegativity
  - Typically, difference in electronegativity is more than 2.0

Ions

Ions are charged atoms or molecules. Ions are formed when atoms or groups of atoms gain or lose valence electrons.

- Monatomic ion – single atom with more or less electrons than the number of electrons in the atom's neutral state
- Polyatomic ions – group of atoms with excess or deficient number of electrons
- Anion – negatively charged ion
- Cation – positively charged ion
- Ionic compounds – association of a cation and an anion
  - The cation is always named first

Nomenclature

Rules for Charges on Monoatomic Ions

- Elements in group 1 form monoatomic ions with charges equal to their group number
  - Na is a group one element, forms Na^+, +1 charge
- Elements in group 2 form monoatomic ions with charges equal to their group number
  - Mg is a group two element, forms Mg^{2+}, +2 charge
- Elements in group 17 form monoatomic ions with a -1 charge
  - Example: Cl^-, F^-, I^-

Cations

- Monatomic cations are formed from metallic elements
  - Na^+ - sodium ion
  - Zn^{2+} - zinc ion

- Some elements can form more than one cation
  - The charge on the ion is indicated by a Roman numeral in parentheses followed by the name of the metal
    - Fe^{2+} - iron (II) ion
    - Fe^{3+} - iron (II) ion
  - Transition metals often form two or more different monoatomic cations

Anions

- Monoatomic anions are typically formed from nonmetals
  - Named by dropping the element name ending and adding –ide
    - Cl^- - chloride ion
    - F^- - fluoride ion
- Common polyatomic anions
  - OH – hydroxide ion
  - CN – cyanide ion
- Many polyatomic anions contain oxygen, they are called oxyanions
  - In elements that form two different oxyanions, the name of the one that contains more oxygen ends in -ate, the one with less ends in -ite:
    - NO_2^- - nitrite ion
    - NO_3^- - nitrate ion
  - Some compounds have multiple oxyanion forms
    - ClO^- - hypochlorite ion, prefix "hypo" added to the oxyanion with the least number of oxygen, suffix "-ite"
    - ClO_2^- - chlorite ion
    - ClO_3^- - chlorate ion
    - ClO_4^- - perchlorate ion, prefix "per" added to the oxyanion with the highest number of oxygen, suffix "-ate"

- Many polyatomic anions with high (negative) charges can add one or more hydrogen cations (H+) to form anions with lower negative charge, their naming reflects whether the H+ addition involves one or more hydrogen ions

    - HSO_4^- - hydrogen sulfate ion

    - $H_2PO_4^-$ - dihydrogen phosphate ion

Acids

According to the Bronsted-Lowery definition, an acid is proton donating (donates H^+).

- Anions with names ending in -ide have associated acids that have the hydro-prefix and an -ic suffix:

    - Cl^- - chloride ion

        - HCl – *hydro*chlor*ic* acid

- Acids of oxyanions

    - If the oxyanion has an -ate ending, the corresponding acid is given an -ic ending

    - If the oxyanion has an -ite ending, the corresponding acid has an -ous ending

    - Prefixes used in the naming of the anion are kept in the name of the acid

        - ClO^- - hypochlorite ion, HClO - hypochlorous acid

        - ClO_2^- - chlorite ion, $HClO_2$ - chlorous acid

        - ClO_3^- - chlorate ion, $HClO_3$ chloric acid

        - ClO_4^- - perchlorate ion, $HClO_4$ - perchloric acid

Molecular Compounds

- A pair of elements can form several different molecular compounds

    - Prefixes are used to identify the relative number of atoms in these compounds

        - CO carbon **mono**xide

        - CO_2 carbon **di**oxide

• Prefixes

1: mono-	5: penta-	9: nona-
2: di-	6: hexa-	10: deca-
3: tri-	7: hepta-	11: undeca-
4: tetra-	8: octa-	12: dodeca-

Hydrates

- Hydrates are compounds that contains water molecules chemically bound to another compound or element

- Hydrates are first named from the anhydrous (dry) compound

  o It is then followed by the word "hydrate" and a prefix to indicate the number of water molecules

  - $CuSO_4 \cdot 5\ H_2O$ – copper (II) sulfate **pentahydrate**

Monoatomic Cations and Anions

Common Cations			Transition Element Cations			Anions		
Charge	Formula	Name	Charge	Formula	Name	Charge	Formula	Name
+1	H^+	Hydrogen	+2	Cd^{2+}	Cadmium	-1	H^-	Hydride
+1	Li^+	Lithium	+2	Cr^{2+}	Chromium(II)	-1	F^-	Fluoride
+1	Na^+	Sodium	+3	Cr^{3+}	Chromium(III)	-1	Cl^-	Chloride
+1	K^+	Potassium	+2	Mn^{2+}	Manganese (II)	-1	Br^-	Bromide
+1	Cs+	Cesium	+2	Fe^{2+}	Iron(II)	-1	I^-	Iodide
+2	Mg^{2+}	Magnesium	+3	Fe^{3+}	Iron(III)	-2	O_2^-	Oxide
+2	Ca^{2-}	Calcium	+2	Co^{2+}	Cobalt(II)	-2	S_2^-	Sulfide
+2	Sr^{2-}	Strontium	+3	Co^{3+}	Cobalt(III)			
+2	Ba^{2-}	Barium	+2	Ni^{2-}	Nickel(II)			
+3	Al^{3+}	Aluminum	+1	Cu^+	Copper(I)			
			+2	Cu^{2+}	Copper(II)			
			+2	Zn^{2+}	Zinc			
			+1	Hg_2^{2+}	Mercury(I)			
			+2	Hg^{2-}	Mercury(II)			

Name	Formula	Name	Formula
Mercury(I) or mercurous	Hg_2^{2+}	Nitrite	NO_2^-
Ammonium	NH_4^+	Nitrate	NO_3^-
Cyanide	CN^-	Hydroxide	OH^-
Carbonate	CO_3^{2-}	Peroxide	O_2^{2-}
Hydrogen carbonate (or bicarbonate)	HCO_3^-	Phosphate	PO_4^{3-}
		Monohydrogen phosphate	HPO_4^{2-}
Acetate	$C_2H_3O_2^-$	Dihydrogen phosphate	$H_2PO_4^-$
Oxalate	$C_2O_4^{2-}$	Sulfite	SO_3^{2-}
Hypochlorite	ClO^-	Sulfate	SO_4^{2-}
Chlorite	ClO_2^-	Hydrogen sulfite (or bisulfite)	HSO_3^-
Chlorate	ClO_3^-		
Perchlorate	ClO_4^-	Hydrogen sulfate (or bisulfate)	HSO_4^-
Chromate	CrO_4^{2-}		
Dichromate	$Cr_2O_7^{2-}$	Thiosulfate	$S_2O_3^{2-}$
Permanganate	MnO_4^-		

More Polyatomic Ions

Formula	Name	Formula	Name
Cations		**Cations**	
NH_4^+	**ammonium**	NH_4^+	**ammonium**
H_3O^+	**hydronium**	H_3O^+	**hydronium**
Anions		**Anions**	
CH_3COO^- (or $C_2H_3O_2^-$)	**acetate**	CrO_4^-	chromate
CN^-	cyanide	$Cr_2O_7^{2-}$	**dichromate**
OH^-	**hydroxide**	O_2^2	peroxide
ClO	hypochlorite	PO_4^{3-}	**phosphate**
ClO_2	chlorite	HPO_4^2	hydrogen phosphate
ClO_3^-	**chlorate**		
ClO_4^-	**perchlorate**	H_2PO_4	dihydrogen phosphate
NO_2	nitrite		
NO_3^-	nitrate	SO_3^{2-}	sulfite
MnO_4^-	**permanganate**	SO_4^{2-}	**sulfate**
CO_3^{2-}	**carbonate**	HSO_4	hydrogen sulfate (or bisulfate)
HCO_3^-	**hydrogen carbonate (or bicarbonate)**		

Oxoanion		Oxoacid	
CO_3^{2-}	Carbonate ion	H_2CO_3	Carbonic acid
NO_2^-	Nitrite ion	HNO_2	Nitrous acid
NO_3^-	Nitrate ion	HNO_3	Nitric acid
PO_4^{3-}	Phosphate ion	H_3PO_4	Phosphoric acid
SO_3^{2-}	Sulfite ion	H_2SO_3	Sulfurous acid
SO_4^{2-}	Sulfate ion	H_2SO_4	Sulfuric acid
ClO^-	Hypochlorite ion	$HClO$	Hypochlorous acid
ClO_2^-	Chlorite ion	$HClO_2$	Chlorous acid
ClO_3^-	Chlorate ion	$HClO_3$	Chloric acid
ClO_4^-	Perchlorate ion	$HClO_4$	Perchloric acid

Chemical Equations

- Chemical reactions are expressed through chemical equations
- An arrow ("→") in a chemical equation means "yields"
    - $2 H_2(g) + O_2(g) → 2 H_2O(l)$
        - Hydrogen + oxygen yields water
    - H_2 and O_2 are reactants
        - Substances that undergo change during a reaction
    - H_2O is the product
        - Substances formed from chemical reactions
- Common phase notation
    - g = gas
    - l = liquid
    - s = solid

Balancing Chemical Equations

- Balanced chemical equations adhere to the Law of Conservation of Matter

    o A balanced equation has to have equal numbers of each type of atom on both sides of the arrow

- Balancing is done by changing the coefficients

    o The coefficient times the subscript gives the total number of atoms

    o If there are no coefficients in front, coefficient is equal to one

    o If an atom doesn't have a subscript, subscript is equal to one

- Subscripts are **never** changed

CHAPTER 3 – STOICHIOMETRY OF FORMULAS AND EQUATIONS

Mass and Moles

In the metric system, the standard unit of mass is the gram (or kilogram).

- All elements have a unique mass (atomic weight)

    - Expressed as either atomic mass units (amu) or grams

    - Same weight of two different elements represents a different number of atoms

- Consider the reaction: $H_2 + F_2 \rightarrow 2\ HF$

    - Does not mean that 1 gram of hydrogen will react with 1 gram of fluorine to form 2 grams of hydrogen fluoride

    - In reality 2.016 g of hydrogen will react with 38.000 g of fluorine to form 40.016 g hydrogen fluoride

        - 2.016 g of hydrogen contain the same number of H_2 molecules as 38.000 g of fluorine (F_2)

        - 40.016 grams of HF will contain twice as many molecules

- Number of molecules, even in low masses, are extremely large numbers

    - So for convenience, amounts in chemistry are expressed in moles

- Mole - quantity of a substance that contains the same number of atoms, molecules or formula units as exactly 12 g of carbon-12

    - 1 mole (mol) = 6.0221×10^{23}

- Atomic mass – mass of one molecule

    - Expressed in atomic mass units (amu)

- Molar mass – mass of one mole of entities (atoms, molecules, formula units) of a substance

    - Expressed in g/mole

- Molar mass and atomic mass are numerically similar
  - Example: one molecule of carbon has an atomic mass of 12.0107 amu and a molar mass of 12.0107 g/mol
    - In 12.0107 g of carbon there are 6.0221×10^{23} molecules

Mass Percentage (Percent Composition)

Mass percentage is a way of expressing the concentration of an element in a compound or a compound of a mixture. Steps for solving percent composition (aka mass percentage) questions:

Example Question: Find the mass percentages of C, O, and H in glucose ($C_6H_{12}O_6$)

- First, look up the atomic masses of the elements that are in the compound on a periodic table
  - C – 12.01 g
  - H – 1.01 g
  - O – 16.00 g
- Second, determine how many grams of each element are in one mole of glucose (or whatever compound a question may be asking you for)
  - C – (6 moles of C x 12.01 g) = 72.06 g
  - H – (12 moles of H x 1.01 g) = 12.12 g
  - O – (6 moles of O x 16.00 g) = 96.00 g
- Third, determine the total mass in one mole of the compound by adding up the masses of the elements from step 2
  - Mass of one mole of glucose = 180.18g (72.06 g + 12.12 g + 96.00 g)
- Finally, find the mass percentages of the elements by dividing the weight of each element in one mole of the compound by the molar mass of that compound
  - C – (72.06 g / 180.18) x 100% = 39.99%
  - H – (12.12 g / 180.18) x 100% = 6.73%
  - O – (96.00 g/ 180.18) x 100% = 53.28%
- To check your work you can add up the percentages to see if they add up to 100%
  - 39.99% + 6.73% + 53.28% = 100%

Formula

$$Mass \ \% \ of \ Element \ Z = \frac{moles \ of \ Z \ in \ formula \ * \ molar \ mass \ of \ Z \ \left(\frac{g}{mol}\right)}{mass \ (g) \ of \ 1 \ mole \ of \ compound}$$

Determining Empirical Formula

Empirical formulas are the smallest whole number ratio of numbers of the atoms in a molecule. The molecular formula of a compound is the formula of the compound as it exists, and may be a multiple of the empirical formula.

Determining Empirical Formula from Masses

Example Question: A compound contains 36.42 g of carbon, 6.12 g of hydrogen, and 47.89 g of oxygen, what is its empirical formula?

- First, determine the moles of each element

    - C – (36.42 / 12.01) = 3.03

    - H – (6.12 / 1.01) = 6.06

    - O – (47.89 / 16.00) = 2.99

- Second, determine the lowest whole-number ratios; divide the moles of each element by the lowest mole amount

    - C – (3.03 / 2.99) = 1.01 → 1

    - H – (6.06 / 2.99) = 2.03 → 2

    - O – (2.99 / 2.99) = 1.00 → 1

        - This step usually results in ratios that are very close to a whole number

        - However, in some question you may get ratios of 1.5, or 2.5, or 3.5, etc. in this case you would multiply all the ratios by 2 to get whole number ratios

        - In some question you may get ratios of 1.33, or 2.33, or 3.33, etc. in this case you would multiply all the ratios by 3 to get whole number ratios

        - In general terms, if the ratios are not very close to a whole number you have to multiply them by a number that would result in approximately whole numbers

- Write the empirical formula from the results

    - CH_2O

Determining Empirical Formula from Elemental Analysis (% Composition)

Example Question: A compound is found to contain 56% carbon, 7% hydrogen, and 37% oxygen. What is the empirical formula for this compound? The molecular weight for this compound is 86.14 g/mol. What is the molecular formula?

- First, assume exactly 100 g of the compound is present
    - This allows you to exchange percentages with grams
        - C – 56% → 56 g
        - H – 7% → 7 g
        - O – 37% → 37 g
- Second, convert masses to moles
    - C – (56 / 12.01) = 4.66 moles
    - H – (7 / 1.01) = 6.93 moles
    - O – (37 / 16.00) = 2.31 moles
- Third, determine the lowest whole-number ratios; divide the moles of each element by the lowest mole amount
    - C – (4.66 / 2.31) = 2.02 → 2
    - H – (6.93 / 2.31) = 3.00 → 3
    - O – (2.31 / 2.31) = 1.00 → 1
- Write the empirical formula from the results
    - C_2H_3O
- To determine the molecular formula from the empirical formula follow these steps:
    - Calculate the weight from the empirical formula (multiply atoms of each element with the elements molar mass and add them up)
        - 2 carbon atoms x 12.01 g = 24.02 g
        - 3 hydrogen atoms x 1.01 g = 3.03 g
        - 1 oxygen atom x 16.00 g = 16.00 g
        - Total : 24.02 g + 3.03 g + 16.00 g = 43.05 g

- Divide the molecular weight by the weight determined from the empirical formula to find the scaling factor

  - 86.14 / 43.05 = 2.00

  - Scaling factor is 2

- Using the scaling factor determine the molecular formula

  - $C_4H_6O_2$

Stoichiometry

Stoichiometry involves using relationships between elements, compounds, chemical formulas, and chemical reactions to acquire quantitative data. There are four major categories of stoichiometry problems that you are likely to encounter. They are listed below with strategies on how to solve them.

- To convert from the mass of a substance to moles of that substance you divide by the molar mas

- To convert from moles of a substance to the mass of a substance you multiply by the molar mass

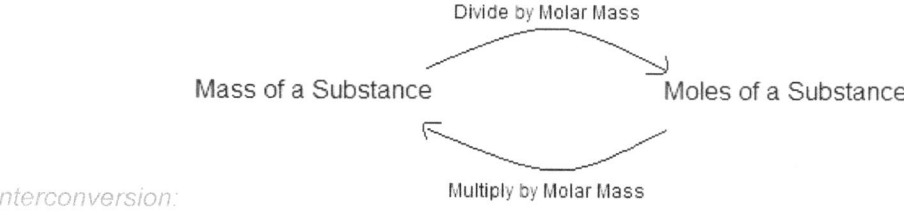

Divide by Molar Mass

Mass of a Substance   Moles of a Substance

Interconversion: Multiply by Molar Mass

- This interconversion is very important in chemical calculations

Stoichiometric Mole–Mole Problems

Example Question: How many moles of HCl are needed to react with 0.82 moles of Al?

- Write out a chemical equation from the information given in the question

  - $Al + HCl \rightarrow AlCl_3 + H_2$

- Balance the chemical equation

  - $2\ Al + 6\ HCl \rightarrow 2\ AlCl_3 + 3\ H_2$

- Calculate the moles of the substance you are told to find using mole ratios

  - $\dfrac{0.82\ mol\ Al}{}\ \Big|\ \dfrac{6\ mol\ HCl}{2\ mol\ Al}\ = \boxed{2.46\ mol\ HCl}$

Stoichiometric Mass–Mass Problems

Example Question: How many grams of Al can be created from decomposing 12.60 g of Al_2O_3?

- Write out a chemical equation from the information given in the question

  o $Al_2O_3 \rightarrow Al + O_2$

- Balance the chemical equation

  o $2\ Al_2O_3 \rightarrow 4\ Al + 3\ O_2$

- Convert the mass of the given substance into moles of the given (mol/g)

  o $\dfrac{12.6\ g\ Al_2O_3}{}\bigg|\dfrac{1\ mol\ Al_2O_3}{101.96\ g\ Al_2O_3} = \boxed{0.12\ mol\ Al_2O_3}$

- Calculate the moles of the substance you are trying to find the mass for using mole ratios

  o $\dfrac{0.12\ mol\ Al_2O_3}{}\bigg|\dfrac{4\ mol\ Al}{2\ mol\ Al_2O_3} = \boxed{0.24\ mol\ Al}$

- Calculate the mass of the substance you are told to find using the moles of the substance calculated in the previous step

  o $\dfrac{0.24\ mol\ Al}{}\bigg|\dfrac{26.98\ g\ Al}{1\ mol\ Al} = \boxed{6.48\ g\ Al}$

Stoichiometric Mass-Volume Problems

Example Question: How many liters of H_2 are created from the reaction of 40.00 g of K in water?

Important: 1 mole of any ideal gas will occupy 22.41 L

- Write out a chemical equation from the information given in the question

  o $K + H_2O \rightarrow KOH + H_2$

- Balance the chemical equation

  o $2\ K + 2\ H_2O \rightarrow 2\ KOH + H_2$

- Convert the mass of the given substance into moles of the given (mol/g)

  - $$\frac{40.0\ g\ K}{}\ \bigg|\ \frac{1\ mol\ K}{39.10\ g\ K} = \boxed{1.02\ mol\ K}$$

- Calculate the moles of the substance you are trying to find the volume for using mole ratios

  - $$\frac{1.02\ mol\ K}{}\ \bigg|\ \frac{1\ mol\ H_2}{2\ mol\ K} = \boxed{0.51\ mol\ H_2}$$

- Calculate the volume of the substance you are told to find using the moles of the substance calculated in the previous step

  - $$\frac{0.51\ mol\ H_2}{}\ \bigg|\ \frac{22.41\ L\ H_2}{1\ mol\ H_2} = \boxed{11.43\ L\ H_2}$$

Stoichiometric Volume-Volume Problems

Example Question: How many liters of SO_2 will be produced from 28.3 L O_2?

Important: 1 mole of any ideal gas will occupy 22.41 L

- Write out a chemical equation from the information given in the question

  - $S_2 + O_2 \rightarrow SO_2$

- Balance the chemical equation

  - $S_2 + 2\ O_2 \rightarrow 2\ SO_2$

- Convert the volume of the given substance into moles of the given (mol/g)

  - $$\frac{28.3\ L\ O_2}{}\ \bigg|\ \frac{1\ mol\ O_2}{22.41\ L\ O_2} = \boxed{1.26\ mol\ O_2}$$

- Calculate the moles of the substance you are trying to find the volume for using mole ratios

  - $$\frac{1.26\ mol\ O_2}{}\ \bigg|\ \frac{2\ mol\ SO_2}{2\ mol\ O_2} = \boxed{1.26\ mol\ SO_2}$$

- Calculate the volume of the substance you are told to find using the moles of the substance calculated in the previous step

  - $$\frac{1.26\ mol\ SO_2}{}\ \bigg|\ \frac{22.41\ L\ SO_2}{1\ mol\ SO_2} = \boxed{28.24\ L\ SO_2}$$

Limiting Reagent

- Typically one of the reactants in a chemical reaction is present in smaller stoichiometric amounts than the other reactant(s)

    o This limits the amount of product(s) that can form

- Limiting reagent problems are easy to identify

    o The problem will give the amounts of more than one of the starting materials

    o You will have to determine the limiting reagent from calculations in order to solve the problem

- There are two typical methods of determining the limiting reagent

    o Need vs. Have Method

    o Product Method

Need vs. Have Method

- Pick one of the reactants (Reac1), calculate how much of the other reactant (Reac2) you will need to completely react with Reac1

    o Utilize stoichiometric ratios

        ▪ This requires a balanced equation!

        ▪ Make sure the equation is balanced before you begin working on the problem

- Compare the amount needed for the Reac2 with the actual amount listed in the question

    o If the amount you need is MORE than what you actually have available according to the problem then Reac2 is the limiting reagent

    o If the amount you need is LESS than what you actually have available according to the problem then Reac1 is the limiting reagent

- You finish the rest of the problem based on the reactant you found to be limiting

"Product" Method

- Pick one of the products of the reaction

    - Choose one of the reactants (Reac1) and calculate how much of the product you can make using the amount you have available according to the question

    - Make the same calculation using the other reactant (Reac2)

- The reactant that results in the smallest amount of product is the limiting reagent

    - The amount is also the maximum amount of product that can be made

        - This amount is called the "theoretical yield"

Reagent in Excess

- Some questions may ask you to determine how much of the excess reagent will be leftover once the reaction ends

- To calculate this value

    - Calculate how much of the excess reagent is needed to completely react with the limiting reagent

    - Take the difference (subtract) off the amount of excess reagent you had at the beginning and the amount needed to completely react with the limiting reagent

Percent Yield

- It is very common that the amount of product made experimentally is lower than the amount expected by the theoretical yield

    - This occurs because of mechanical errors, incomplete reactions, "side" reactions, etc.

- To determine the percent yield (sometimes also called the efficiency of a reaction) use the following formula:

    - $\% \, Yield = \frac{actual \, yield}{theoretical \, yield} \; x \; 100\%$

CHAPTER 4 – CHEMICAL REACTIONS

Water as a Solvent

- Most reactions that occur in organisms and the environment take place in water

- Water has 2 hydrogen atoms covalently bonded to 1 oxygen atom

- The electronic structure of water is tetrahedral

    - 2 covalent bonds with H atoms and 2 sets of unpaired electrons

    - 104.5° bond angle – because the two lone pairs try to separate as far as possible

- Water is a **polar** molecule

    - Polar atoms have dipoles because of unequal sharing of electrons

        - The shared pair of electrons are attracted more strongly to the oxygen forming a partially negative charged "pole" near the oxygen and partially positively charged poles near the Hydrogens

Structure of Water with Dipoles: $\delta^+H - \overset{\delta^-}{O} - H\delta^+$

Solubility

- Soluble substances (solutes) in water

    - Interaction between an ion and water is strong

    - Polar compounds are water soluble

        - Polar water molecules have an affinity for oppositely charged regions of other polar molecules

- Insoluble substances (solutes) in water

    - Non-polar compounds are not water-soluble

    - If the interaction between ion and water is weak

        - Insoluble substances have a force of attraction so strong in their solid substance form, that it cannot be overcome by the interaction of the ions with the polarized water molecules

Ions in Aqueous Solution

- In ionic solids, cations & anions are held together by electrostatic attraction

    o The electrostatic attraction is replaced by water molecules in an aqueous solution

- Many ionic compounds dissociate into independent ions when dissolved in water

    o Compounds that freely dissociate into independent ions in aqueous solution are called electrolytes

        ▪ Solutions with electrolytes are good conductors of electricity

        ▪ Not all electrolytes are ionic compounds

- Some molecular compounds dissolve but do not dissociate into ions

    o Contain polar bonds which interact with the polar bonds in water

    o Do not conduct an electric current

        ▪ Produce a nonconducting solution

    o Referred to as nonelectrolytes

Strong and Weak Electrolytes

- Strong electrolyte - an electrolyte that completely disassociates in solution

    o Exist in solution almost entirely as ions

- Weak electrolyte - an electrolyte that does not completely disassociate in solution

    o Exists in solution as both ions and molecules of the electrolyte

Molecular and Ionic Equations

- Molecular equation express reactants and products as if they were molecules, despite actually existing in solution as ions

    o Example of a molecular equation

        ▪ $CaCl_2 \text{ (aq)} + 2 \, AgNO_3 \text{ (aq)} \rightleftharpoons Ca(NO_3)_2 \text{ (aq)} + 2 \, AgCl \text{ (s)}$

        ▪ (aq) is used to indicate that the substance is actually disassociated in solution

        ▪ (s) is used to indicate that the substance is a precipitate

- Ionic equation express strong electrolytes as separate independent ions

  o Example of an ionic equation

    ▪ $Ca^{2+}_{(aq)} + 2\ Cl^-_{(aq)} + 2\ Ag^+_{(aq)} + 2\ NO_3^-_{(aq)} \rightleftharpoons Ca^{2+}_{(aq)} + 2\ (NO_3)^-_{(aq)} + 2\ AgCl_{(s)}$

    ▪ Note that the solid is written in its full formula

- Net ionic equation are equations without any spectator ions

  o Spectator ion - an ion in an ionic equation that does not take part in the reaction

  o Easy way of identifying spectator ions is by writing the ionic equation and then crossing off any ions that appear on both sides of the equation (those are the spectator ions)

  o Example:

    ▪ $\cancel{Ca^{2+}_{(aq)}} + 2\ Cl^-_{(aq)} + 2\ Ag^+_{(aq)} + \cancel{2\ NO_3^-_{(aq)}} \rightleftharpoons \cancel{Ca^{2+}_{(aq)} + 2\ (NO_3)^-_{(aq)}} + 2\ AgCl_{(s)}$

    ▪ Resulting net equation: $2\ Cl^-_{(aq)} + 2\ Ag^+_{(aq)} \rightleftharpoons 2\ AgCl_{(s)}$

Three Major Classes of Chemical Reactions

The three major classes of chemical reactions are: precipitation reactions, acid-base reactions, and oxidation-reduction reactions.

Precipitation Reactions

- Precipitation reaction occurs in aqueous solution because one of the product in a precipitation reaction is insoluble

  o Precipitate - an insoluble solid compound formed during a chemical reaction in solution

- There are generalized solubility rules that are used to predict whether a precipitate will form

  o Soluble - all compounds containing the ammonium ion (NH_4^+) or alkali metal (Group IA on the periodic table) cations

  o Soluble - all nitrates and acetates (ethanoates)

  o Soluble - all chlorides, bromides and iodides

    ▪ Except those with Ag, Pb, Hg

- ○ Soluble - all sulfates

  - ▪ Except those with Ag, Pb, Hg(I), Ba, Sr, Ca

- ○ Insoluble - all carbonates, sulfites, and phosphates

  - ▪ Except those with ammonium (NH_4^+), and alkali metal (Group IA) cations

- ○ Insoluble - all Hydroxides

  - ▪ Except those with NH_4^+, alkali metal (Group IA) cations

- ○ Insoluble - all sulfides

  - ▪ Except those with NH_4^+, alkali metal (Group Ia) cations, and alkali earth metal (Group II) cations

- ○ Insoluble - all oxides

  - ▪ Except those with Calcium, Barium, and alkali metal (group IA) cations

Solubility Rules Tables:

Solubility Rules for Ionic Compounds

Rule	Applies to	Statement	Exceptions
1	Li^+, Na^+, K^+, NH_4^+	Group IA and ammonium compounds are soluble.	—
2	$C_2H_3O_2^-$, NO_3^-, ClO_4^-	Acetates and nitrates are soluble.	—
3	Cl^-, Br^-, I^-	Most chlorides, bromides, and iodides are soluble.	$AgCl$, Hg_2Cl_2, $PbCl_2$, $AgBr$, $HgBr_2$, Hg_2Br_2, $PbBr_2$, AgI, HgI_2, Hg_2I_2, PbI_2
4	SO_4^{2-}	Most sulfates are soluble.	$CaSO_4$, $SrSO_4$, $BaSO_4$, Ag_2SO_4, Hg_2SO_4, $PbSO_4$
5	CO_3^{2-}	Most carbonates are insoluble.	Group IA carbonates, $(NH_4)_2CO_3$
6	PO_4^{3-}	Most phosphates are insoluble.	Group IA phosphates, $(NH_4)_3PO_4$
7	S^{2-}	Most sulfides are insoluble.	Group IA sulfides, $(NH_4)_2S$
8	OH^-	Most hydroxides are insoluble.	Group IA hydroxides, $Ca(OH)_2$, $Sr(OH)_2$, $Ba(OH)_2$

Negative Ions (Anions)	+	Positive Ions (Cations)	=	Solubility of Compounds In Water	Example
Any Anion	+	Alkali Ions $(Li^+, Na^+, K^+, Rb^+, Cs^+, Fr^+)$	=	Soluble	NaF, KNO_3
Any Anion	+	Hydrogen Ion $[H^+(aq)]$	=	Soluble	HCl
Any Anion	+	Ammonium Ion (NH_4^+)	=	Soluble	NH_4Cl
Nitrate (NO_3^-)	+	Any Cation	=	Soluble	$Ca(NO_3)_2$
Acetate (CH_3COO^-)	+	Any Cation	=	Soluble	CH_3COONa
Halides (Cl^-, Br^-, I^-)	+	$Ag^+, Pb^{2+}, Hg^{2+}, Cu^+, Tl^+$	=	Insoluble	$AgCl$, $PbBr_2$
	+	Any Other Cation	=	Soluble	KBr, CaI_2
Sulfate (SO_4^{2-})	+	$Ca^{2+}, Sr^{2+}, Ba^{2+}, Ag^+, Pb^{2+}, Ra^{2+}$	=	Insoluble	$BaSO_4$
	+	Any Other Cation	=	Soluble	$CuSO_4$
Sulfide (S^{2-})	+	Alkali Ions – $Li^+, Na^+, K^+, Rb^+, Cs^+, Fr^+$ Alkali Earth Metals – $Be^{2+}, Mg^{2+}, Ca^{2+}, Sr^{2+}, Ba^{2+}, Ra^{2+}, H^+(aq), NH_4^+$	=	Soluble	H_2S, MgS, $(NH_4)_2S$
	+	Any Other Cation	=	Insoluble	ZnS
Hydroxide (OH^-)	+	Alkali Ions - $Li^+, Na^+, K^+, Rb^+, Cs^+, Fr^+, H^+(aq), NH_4^+, Sr^{2+}, Ba^{2+}, Ra^{2+}, Tl^+$	=	Soluble	$Sr(OH)_2$
	+	Any Other Cation	=	Insoluble	AgOH
Phosphate (PO_4^{3-}) Carbonate (CO_3^{2-})	+	Alkali Ions - $Li^+, Na^+, K^+, Rb^+, Cs^+, Fr^+, H^+(aq), NH_4^+$	=	Soluble	$(NH_4)_3PO_4$
	+	Any Other Cation	=	Insoluble	$MgCO_3$

- Precipitation reactions take the form of an "exchange reaction"

  o Exchange reactions are reaction between compounds that appear to involve an exchange of cations and anions

Acid-Base Reactions

- The Arrhenius definition of acids and bases

  o Acids produce hydrogen ions (H^+) when dissolved in water

  o Bases produce hydroxide ions (OH^-) when dissolved in water

- Bronsted-Lowery definition of acids and bases

  - Acids are proton donating

  - Bases are proton accepting

- Strong acid – an acid that ionizes completely in water

  - Strong acids are HI, HBr, $HClO_4$, HCl, $HClO_3$, H_2SO_4, and HNO_3

- Weak acid – an acid that only partially ionizes in water

- Strong base – a base that is present almost entirely as ions (one of the ions is OH^-)

  - Strong bases are $NaOH$, KOH, $LiOH$, $RbOH$, $CsOH$, $Ca(OH)_2$, $Ba(OH)_2$, and $Sr(OH)_2$

- Weak base – a base that only partially ionizes in water

- Strong acids and bases are represented as separate ions in an ionic equation

- Weak acids and bases are represented as undissociated "molecules" in ionic equations

  - Example of Acetic Acid (a weak acid)

  - Molecular Equation: $CH_3COOH_{(aq)} + NaOH_{(aq)} \rightarrow CH_3COONa_{(aq)} + H_2O_{(l)}$

  - Ionic: $CH_3COOH_{(aq)} + Na^+_{(aq)} + OH^-_{(aq)} \rightarrow CH_3COO^-_{(aq)} + Na^+_{(aq)} + H_2O_{(l)}$

  - Net: $CH_3COOH_{(aq)} + OH^-_{(aq)} \rightarrow CH_3COO^-_{(aq)} + H_2O_{(l)}$

Neutralization (Acid-Base) Reactions

- Acids and bases neutralize one another

- Neutralization reaction - reaction between an acid and a base that results in an ionic compound and water

- Salt - ionic compound that is the product of a neutralization reaction

- Net ionic equation for each neutralization reaction involves a transfer of a proton

  - Consider the example of HCl (a strong acid) reacting with $LiOH$ (a strong base)

    - Net equation for the reaction is: $H^+_{(aq)} + OH^-_{(aq)} \rightarrow H_2O_{(l)}$

- Consider the example of HCN (a weak acid) reacting with KOH (a weak base)

  - $HCN_{(aq)} + OH^-_{(aq)} \rightarrow CN^-_{(aq)} + H_2O_{(l)}$

  - The proton is transferred from HCN to OH^-

Acid-Base Reactions with Gas Formation (aka Displacement Reactions)

- Acid-base reactions with gas formation sometimes involve unstable chemical species (e.g., H_2CO_3 and H_2SO_3

- Unstable species are enclosed in parentheses

  - Example: $[H_2CO_{3\,(aq)}] \rightarrow H_2O_{(l)} + CO_2\,(g)$

Oxidation-Reduction Reactions (aka Redox Reactions)

- Oxidation-reduction reactions (aka redox reactions) – reactions that involve a partial or complete transfer of electrons from one reactant to another

  - Oxidation = loss of electrons

  - Reduction = gain of electrons

    - Trick for remembering which is which - OIL RIG

    - **OIL - O**xidation **I**s **L**osing electrons

    - **RIG** - **R**eduction **I**s **G**aining electrons

  - Oxidation and reduction always occur simultaneously

- Oxidation number (aka oxidation state) – actual charge an atom in a molecule would have if all the electrons it was sharing were transferred completely, not shared

- Oxidizing agent – species that oxidizes another species

  - It is itself reduced

- Reducing agent – species that reduces another species

  - It is itself oxidized

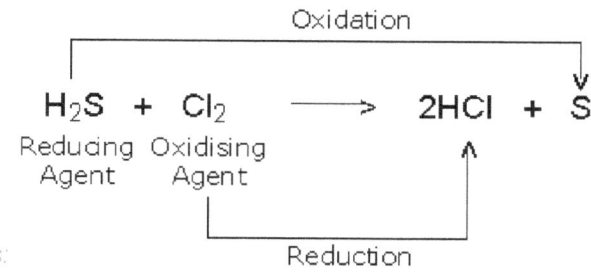

Oxidation Number Rules

Applies to	Rule
Elements	The oxidation number of an atom in an element is zero.
Monatomic ions	The oxidation number of an atom in a monatomic ion equals the charge of the ion.
Oxygen	-2 in oxides, e.g. ZnO, CO_2, H_2O -1 in peroxides, e.g. H_2O_2 $-(1/2)$ in superoxides, e.g. KO_2 $-(1/3)$ in inorganic ozonides, e.g. RbO_3 0 in O_2 $+(1/2)$ in dioxygenyl, e.g. dioxygenyl hexafluoroarsenate $O_2{}^+ [AsF_6]^-$ $+1$ in O_2F_2 $+2$ in OF_2
Hydrogen	The oxidation number of hydrogen is $+1$ in most of its compounds
Halogens	Fluorine is -1 in all its compounds. The other halogens are -1 unless the other element is another halogen or oxygen
Compounds and ions	The sum of the oxidation numbers of the atoms in a neutral compound is zero. The sum in a polyatomic ion equals the charge on the ion
Carbon in Organic Compounds	The oxidation number of Carbon in Organic Compounds has an average value of 0 The O.N. of other elements in carbon compounds (H^{+1}, O^{-2}, N^{+3}, Cl^{-1}, etc.) still apply

- Common Redox Reactions

  o Combination - reaction in which two substance combine to form a third substance

  o Decomposition – reaction in which a single compound produces two or more substances

  o Displacement (aka single-replacement reaction) – reaction in which an element reacts with a compound and displaces an element from it

  o Combustion – reaction in which one of the reactants is oxygen

    ▪ Usually results in the rapid release of heat

- Redox reactions can be written in terms of two half-reactions

  o One involves the loss of electrons (oxidation)

  o The other involves the gain of electrons (reduction)

  o Example: $Fe^{2+} + Ce^{4+} \rightarrow Fe^{3+} + Ce^{3+}$

$$Fe^{2+} \rightarrow Fe^{3+} + e^{-} \quad \text{(oxidation half-reaction)}$$
$$Ce^{4+} + e^{-} \rightarrow Ce^{3+} \quad \text{(reduction half-reaction)}$$
$$\overline{Fe^{2+} + Ce^{4+} \rightarrow Fe^{3+} + Ce^{3+}}$$

- A balanced redox equation has to have charge balance

  o Number of electrons lost in the oxidation half-reaction must be equal to the number of electrons gained in the reduction half-reaction

CHAPTER 5 – QUANTUM THEORY AND ATOMIC STRUCTURE

Emission Spectrum

- When elements are burned in a flame and their emissions are passed through a prism only a few color lines are seen

    - These lines are a distinct characteristic of each element

- Atoms emit light of a characteristic wavelengths when they return from an excited state to their ground state

Emission Spectrum of Hydrogen:

Light

Light is a form of electromagnetic energy that behaves as both a wave and a particle.

Light as a Wave

- Electromagnetic energy travels in rhythmic waves which are disturbances of electric and magnetic fields

- Wavelength (λ) - distance between consecutive crests of electromagnetic waves

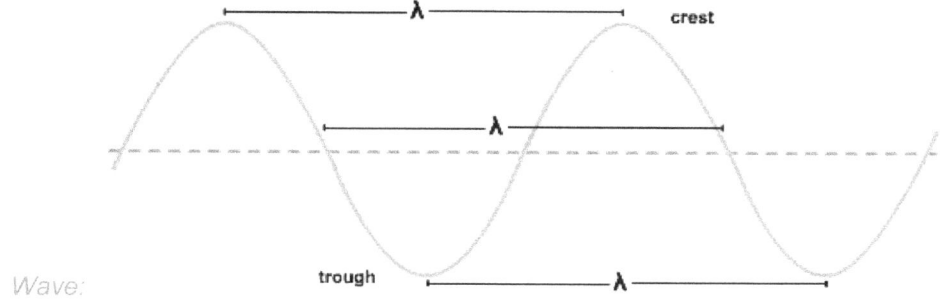

Wave:

- Frequency (f) – number of crests of a wave that move past a given point in a given unit of time

    - $f = \lambda/v$

    - v = speed

- Speed of light in a vacuum = 2.998 x 10^8 m/sec

- The electromagnetic spectrum encompasses a very wide range of wavelengths from as small as a nanometer to those that are more than a kilometer

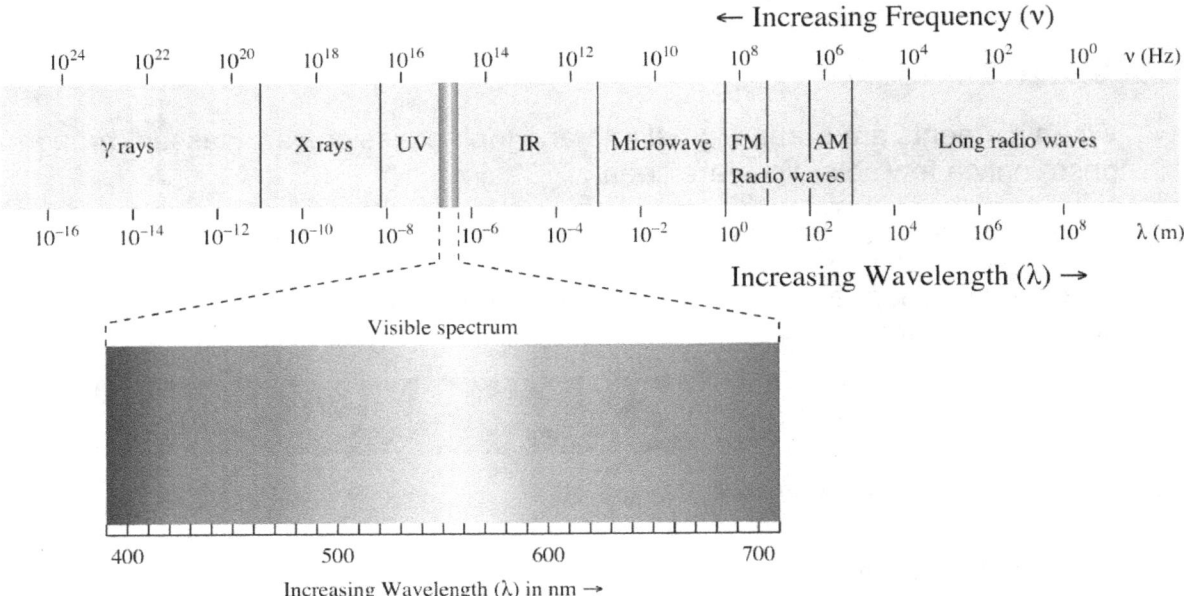

Light as a Particle

- Light also behaves as if it consists of discrete particles or quanta called photons

- Each photon has a fixed quantity of energy

  o E= hc/λ

    ▪ h is the Planck's constant = 6.626 X 10^{-34} joules•sec

    ▪ c is the speed of light in a vacuum = 2.998 X 10^8 m/sec

  o Shorter λ = higher energy

Wave-Particle Duality

- All matter and energy exhibit both wave and particle-like properties

- This duality is seen in:

  o Macroscopic objects

  o Microscopic objects (e.g. atoms and molecules)

  o Quantum objects (e.g., protons, neutrons, quarks, mesons)

- To explain Wave-Particles Duality, physicists focused on 3 phenomena
  - Black Body Radiation
  - Photoelectric Effect
  - Atomic Line Spectra

Black Body Radiation

As the temperature of an object changes, its temperature is directly related to the wavelengths of light that it emits. This is a characteristic of the idealized "black body."

- Max Planck developed a mathematical model to reproduce the spectrum of light emitted by glowing objects
  - The model was developed under the assumption that a vibrating (oscillating) atom can only emit or absorb certain quantities of energy
- Planck's Model
  - $E = nh\nu$
    - E - energy of radiation
    - ν – frequency
    - n – quantum number (1, 2, 3...)
    - h – Planck's Constant = 6.626×10^{-34} J•s
  - Quanta – packets of energy that can be emitted or absorbed
  - Atoms change energy states when they emit or absorb one or more quanta
  - The model views emitted energy as waves

Photoelectric Effect

The photoelectric effect is the observation that many metals emit electrons when light shines upon them. Electrons emitted in this manner are called photoelectrons.

- Electrons are ejected from the surface of the metal only when the frequency exceeds a certain threshold
  - The threshold is dependent on the characteristic of the metal
- Einstein reasoned that atoms change energy states when they emit or absorb a quantum of light energy, which he called a photon
  - He defined a photon as a particle of electromagnetic energy

- $E_{photon} = \Delta E_{atom} = hv = \dfrac{hc}{\lambda}$

- According to the photoelectric effect

    - Electrons exist in different energy states

    - A photon whose frequency is greater than or equal to the energy state of the electron will be absorbed

        - If the frequency is less than the electron energy level, the photon is not absorbed

    - The electron moves to a higher energy state and is ejected from the surface of the metal when it absorbs a photon

    - Electrons are attracted to the positive anode of a battery which causes a flow of current

Atomic Line Spectra

- When light from an "excited" atom passes through a prism, it does not form a continuous spectrum

    - Instead, it produces a series of colored lines called "line spectra" that are separated by black spaces

    - Wavelengths of the lines are a characteristic of the elements producing them

        - Different elements have different line spectra

- Example of Hydrogen

    - The spectra lines of Hydrogen occur in several series

        - Each series is represented by a positive integer, n

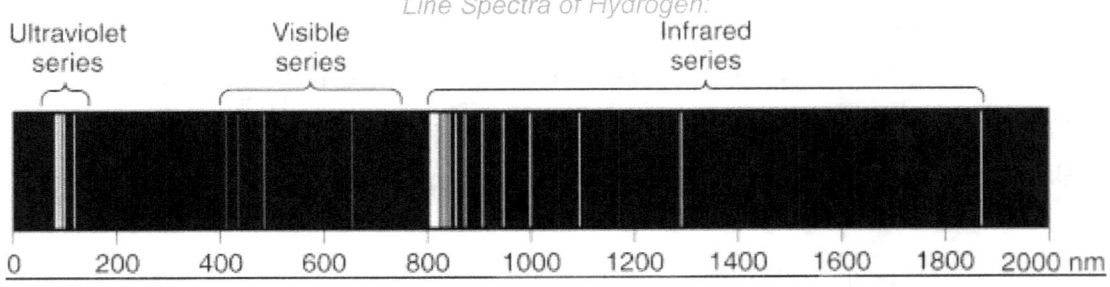

Line Spectra of Hydrogen:

        - Ultraviolet series, n =1

        - Visible series, n = 2

        - Infrared series, n = 3

- o The visible spectrum of Hydrogen can be reproduced by a Rydberg Equation

    - $\frac{1}{\lambda} = R\left(\frac{1}{n_1^2} - \frac{1}{n_2^2}\right) = 1.096776 \times 10^7 \ m^{-1}\left(\frac{1}{n_1^2} - \frac{1}{n_2^2}\right)$

        - R – Rydberg Constant = 1.096776 x 10^7 m^{-1}

        - Λ – wavelength of the spectra line

        - n_1 and n_2 are positive integers where $n_2 > n_1$

<u>Bohr Theory</u>

Bohr's theory of an atom was an early model of atomic structure in which electrons travel around the nucleus in a number of discrete stable orbits determined by quantum conditions.

- Bohr came up with 2 postulates to account for how electrons lose energy yet remain in orbit

    - o A electron has specific energy levels in an atom

    - o A electron in an atom changes energy levels by undergoing a transition from one energy level to another

- Bohr came up with a formula for the energy levels of the electron in the hydrogen atom

    - o E = -2.18 x 10^{-18} J $\left(\frac{Z^2}{n^2}\right)$

        - n – principal quantum numbers = 1, 2, 3…

        - Z – nuclear charge

            - Z = 1 for Hydrogen

- When an electron undergoes a transition from a higher energy level (n_i) to a lower one (n_f) the energy is emitted as a photon

    - o E = hv = E_f – E_i = ΔE

        - E – energy of emitted photon

        - $E_i = -\frac{2.18 \times 10^{-18} \ J}{n_i^2}$

    - o ΔE = hv = -2.18 x 10^{-18} J $\left(\frac{1}{n_f^2} - \frac{1}{n_i^2}\right)$

Quantum Mechanics

Quantum mechanics is the branch of physics that attempts to mathematically describe the wave properties of submicroscopic particles.

- de Broglie relation

  - $\lambda = h/mv$

    - Matter has wave-like properties but these properties are not commonly observed

    - de Broglie relation shows that the wavelength of objects is so incredibly small that these waves cannot be detected

- Heisenberg's Uncertainty Principle

  - States that you can never know the exact position and the exact speed of an object with high precision

    - Essentially, there is an uncertainty

  - The principle is also a relationship that states that the product of the uncertainty in position (Δx) and the uncertainty in momentum ($m\Delta v_x$) of a particle can be no smaller than what is predicted by the equation:

    - $(\Delta x)(m\Delta v_x) \geq \dfrac{h}{4\pi}$

      - $\dfrac{h}{4\pi} = 5.28 \times 10^{-35}$ J•s

    - When m is large (represents a large object) the uncertainty is very small but for subatomic particles like electron there is a high level of uncertainty

      - This uncertainty is why we can't define the exact orbit of an electron

- Schrodinger developed a quantum mechanical model of the hydrogen atom. According to the model:

  - An atom has specific allowed quantities of energy

  - An electron's behavior is wave-like

    - However, its exact location is impossible to know

- The electron's Matter-Wave occupies a 3-D space near the nucleus
  - The Matter-Wave experiences continuous and varying influence from the nuclear charge
- Schrodinger Equation
  - $H^{(op)}\Psi = E\Psi$
    - E – energy of the atom
    - Ψ – wave function
    - $H^{(op)}$ – Hamiltonian Operator

Quantum Numbers and Atomic Orbitals

- Quantum mechanics describes each electron by 4 quantum numbers
  - Principal Quantum Number (n)
  - Angular Momentum Quantum Number (l)
  - Magnetic Quantum Number (m_l)
  - Spin Quantum Number (m_s)
    - n, l, m_l define the wave function of the electron's atomic orbital
    - m_s refers to the spin orientation of the 2 electrons that occupy an atomic orbital
- Principal Quantum Number (n) represents the "shell number" in which an electron is located
  - Represents the relative size of the orbital
  - Defines the principal energy of the electron
  - Smaller "n" represents a smaller orbital and a lower energy of the electron
  - n can take any positive value
- Angular Momentum Quantum Number (l) represents the "sub shells" within a given shell
  - Each main "shell" is designated by a quantum number "n"
    - It is further subdivided into: $l = n - 1$ "sub shells"
    - "l" can have any integer value from 0 to n - 1

- "ℓ" values correspond to the s, p, d, f designations used in the electronic configuration of the elements

  - s has an ℓ value of 0

  - p has an ℓ value of 1

  - d has an ℓ value of 2

  - f has an ℓ value of 3

- Magnetic Quantum Number (m_ℓ) defines atomic orbitals within a given sub-shell

  - Each value of the angular momentum number (ℓ) determines the number of atomic orbitals

  - For any given value of "ℓ," m_l can take any integer value from $-\ell$ to $+\ell$

    - $m_\ell = -\ell$ to $+\ell$

  - Each orbital has a different shape and orientation (x, y, z) in space

  - Each orbital within a given angular momentum number sub shell (ℓ) has the same energy

- Spin Quantum Number (m_s) represents the two possible spin orientations of an electron residing within a given atomic orbital

  - Each atomic orbital holds only two electrons

  - Each electron has a "spin" orientation value

    - These values must be opposite of one another

  - Possible values of m_s are: $+\frac{1}{2}$ and $-\frac{1}{2}$

Summary of Quantum Numbers

Name	Symbol	Permitted Values	Property
Principal	N	positive integers (1, 2, 3...)	orbital energy (size)
Angular Momentum	ℓ	integers from 0 to n-1	orbital shape The ℓ values 0,1, 2, and 3 correspond to s, p, d, and f orbitals, respectively
Magnetic	m_ℓ	integers from $-\ell$ to 0 to $+\ell$	orbital (x, y, z) orientation
Spin	m_s	$+\frac{1}{2}$ or $-\frac{1}{2}$	e⁻ spin orientation

- Shapes of orbitals
  - s (n=1) sub shell orbital
    - Only one orbital, holds 2 electrons, spherical shape
  - p (n=2) sub shell orbitals
    - Three orbitals, holds 6 electrons, dumbbell shape
  - d (n=3) sub shell orbital
    - Five orbitals, holds 10 electrons, pear-shaped lobes and dumbbell shapes
  - f (n=4) sub shell orbitals
    - Seven orbitals, holds 14 electrons

CHAPTER 6 – ELECTRON CONFIGURATION AND PERIODIC PROPERTIES

Brief Review to Understand Electron Configuration

- The quantum numbers n, ℓ, m_ℓ define an orbital

    - A orbital contains a maximum of 2 electrons

        - Each electron has a different spin ($+\frac{1}{2}$ or $-\frac{1}{2}$)

- Orbital diagrams are notations used to show how the orbitals of a sub shell are occupied by electrons

    - Each group of orbitals is labeled by its sub shell notation (s, p, d, f)

    - Electrons are represented by arrows

        - Up ↑ for $m_s = +\frac{1}{2}$

        - Down ↓ for $m_s = -1/2$

- Pauli Exclusion Principle

    - No two electrons in an atom can have the same four quantum numbers

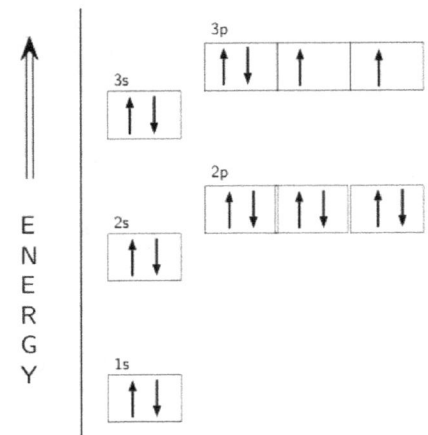

- Summary table for the maximum number of electrons in each sub shell:

Sub Shell	Number of Orbitals	Values for ($-\ell$ to $+\ell$)	Maximum Number of Electrons
s ($\ell =0$)	1	(0)	2
p ($\ell =1$)	3	(-1, 0, +1)	6
d ($\ell =2$)	5	(-2, -1, 0, +1, +2)	10
f ($\ell =3$)	7	(-3, -2, -1, 0, +1, +2, +3)	14

Electron Configuration

The electron configuration of an atom refers to the particular distribution of electrons among the available sub shell in that atom.

- Electronic configuration notation lists subshell symbols (s, p, d, f) sequentially with a superscript to indicate the number of electrons in that subshell

  - Ex. Fluorine

    - Atomic Number: 9

    - Number of electrons in a neutral Fluorine atom: 9

      - Number of electrons for a neutral atom is the same as the atomic number

    - 2 electrons in the "1s" sub shell

    - 2 electrons in the "2s" sub shell

    - 5 electrons in the "2p" sub shell

    - Electron Configuration: $1s^22s^22p^5$

- Configurations can become quite complex as atomic number increases

  - To remedy this, a condensed form of the configuration is often used which utilizes electron configurations of noble gases

    - Noble gases have the maximum number of electrons possible in their outer shell

      - Makes them very unreactive

    - The noble gases are: helium, neon, argon, krypton, xenon, and radon

Table of Condensed Electronic Configuration Examples:

Element	Noble Gas?	Full Electronic Configuration	Condensed Electronic Configuration	Total Number of Electrons
Neon	Yes	$1s^22s^22p^6$	[Ne]	10
Argon	Yes	$1s^22s^22p^63s^23p^6$	[Ar]	18
Krypton	Yes	$1s^22s^22p^63s^23p^63d^{10}4s^24p^6$	[Kr]	36
Beryllium	No	$1s^22s^2$	[He] $2s^2$	4
Magnesium	No	$1s^22s^22p^63s^2$	[Ne] $3s^2$	12
Calcium	No	$1s^22s^22p^63s^23p^64s^2$	[Ar] $4s^2$	20

- o [X] represents the electron configuration of the nearest noble gas that appears before the element of interest on the periodic table
- Keep in mind that you have to adjust the number of electrons and thus the electron configuration for cations and anions of an element

Nuclear Charge, Shielding Effect, and Orbital Shape

Nuclear Charge

- In an atom there are 2 counteracting forces:
  - o In the nucleus
    - Positive protons pull (attract) the negatively charged electrons
  - o Outside the nucleus
    - Negatively charged electrons are repelling each other
- Higher nuclear charge (Z) lowers orbital energy by increasing the amount of proton-electron attractions
  - o Lowering orbital energy makes it more difficult to remove the electron from orbit

Shielding Effect

- The repulsion electrons experience from other electrons shields (counteracts) the attractive force of the protons in the nucleus
- Shielding lowers the full nuclear charge to an "effective nuclear charge" (Z_{eff})
  - o Lowering the effective nuclear charge makes it easier to remove an electron
    - It takes less energy to remove an electron from Helium (He) than from He^+
      - Since the second electron in He repels the first electron and effectively shields the first electron from the full nuclear charge

Effects of Orbital Shape

- Shape of an atomic orbital affects how close an electron comes to the nucleus (i.e. the level of penetration)

- Penetration and shielding cause the energy level (n) to be split into sublevels of differing energy

    o This is represented by the various values of the magnetic quantum number (ℓ)

        ▪ The lower the value of the magnetic quantum number, the greater the electron penetration

    o Order of Sublevel Energies

        ▪ s (ℓ =0) < p(ℓ =1) < d(ℓ =2) < f(ℓ =3)

Aufbau Principle

The Aufbau principle states that electrons orbiting one or more atoms fill the lowest available energy levels before filling higher levels. For example, an electron has to fill the "1s" before "2s."

- Filling orbitals of the lowest energy first gives the lowest total energy of the atom

    o The ground state of the atom

- Order in which the possible sub-shells fill

    o 1s, 2s, 2p, 3s, 3p, 4s, 3d, 4p, 5s, 4d, 5p, 6s, 4f, 5d, 6p, 7s, 5f

    o Order in which subshells are filled is in the order in which the diagonal lines go through the subshells

        ▪ Note: 4s subshell is filled before the 3d subshell because 4s electrons are at a lower energy level than the 3d electrons

        ▪ Lower energy levels are always filled first

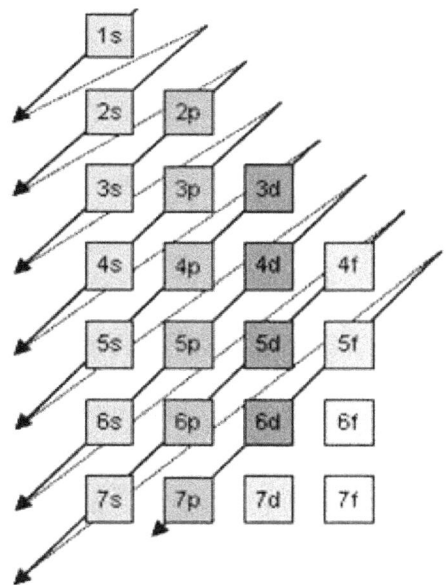

- Configuration associated with the lowest energy level of an atom is the ground state

    o Any other configurations correspond to the excited states

Hund's Rule

Hund's rule states that the "ground state" (i.e. the lowest energy configuration) of electrons in a sub-shell is attained from putting electrons into separate orbitals of a sub shell before pairing the electrons.

Example: Oxygen

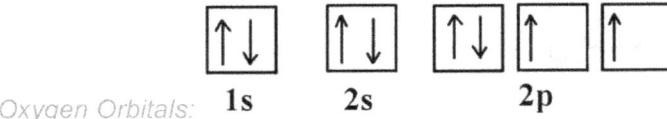

Oxygen Orbitals: 1s   2s   2p

- Note: Two of the electrons in the 2p orbitals appear singly instead of being paired together

    - According to Hund's Rule, electrons occupy separate orbitals rather than pairing when possible

Configurations and the Periodic Table

- Electrons in an atom's outermost shell are called valence electrons

    - Valence electrons are the electrons primarily involved in chemical reactions

    - Elements within a group have the same valence shell configuration

        - This is why groups of elements share chemical properties

- Noble gases have filled outer shells

    - Very high ionization energies

    - Positive (endothermic) electron affinities

    - Do not readily form ions or react

        - Inert for the most part

    - Very stable

        - Other elements try to attain noble gas configuration (filled outer shells) for stability

- Elements in Groups 1A and 2A readily form cations by losing electrons

    - They only have 1 or 2 electrons in their outer shells respectively, losing their valence electrons allows these elements to conform to noble gas configuration

- Elements in Groups 6A and 7A readily form anions by gaining electrons

    o They fill their outer shells and conform to a noble gas configuration

    o They are isoelectronic with the nearest noble gas configuration

        ▪ Isoelectronic – have the same number of electrons or the same electronic structure

- Large metals from Groups 3A, 4A, and 5A form cations through a different process than the smaller 1A and 2A elements

    o They have to lose large amounts of electrons in order to attain noble gas configuration

    o Example: Tin (Sn), would have to lose two electrons from 5p, ten from 4d, two from 5s in order to be isoelectronic with Krypton

        ▪ For Tin it is easier to gain stability by reaching a noble gas-like configuration by having empty 5s & 5p sublevels and a filled inner 4d subshell configuration

- Many times it is only important to know the valence electron number for an atom

    o On the periodic table it is easy to determine this: the number of valence electrons for an atom matches its group number

        ▪ Example: Group 7A elements have 7 valence electrons

Periodic Properties

Periodic Law states that when elements are arranged by atomic number, their physical and chemical properties vary across the periodic table row.

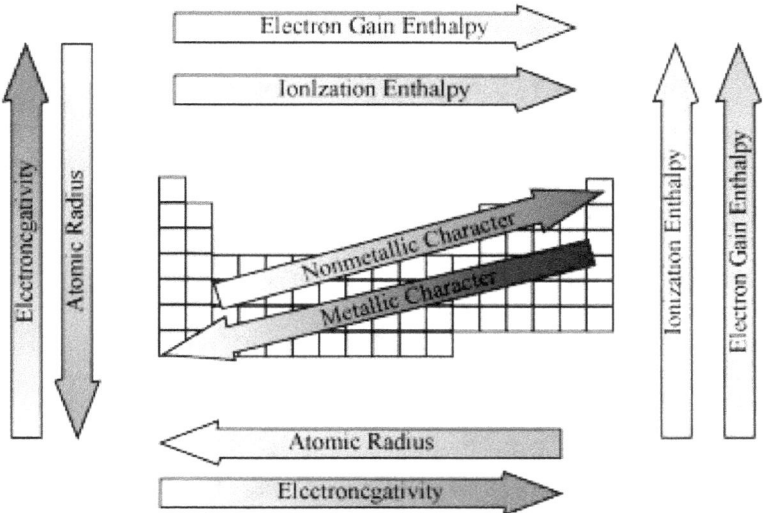

Atomic Size

- Two factors that affect size of an atom
  - Larger the principal quantum number (n), the larger the size of the orbit
  - Effective nuclear charge
    - Positive charge an electron experiences from the nucleus minus any shielding effects
- Atomic radius tends to **decrease** with increasing atomic number across a period
  - There is a higher effective nuclear charge
    - Greater attractive force in the nucleus from the higher number of protons
- Atomic radius tends to **increase** with increasing atomic number within a group
  - More energy levels
    - Each subsequent energy level is further from the nucleus than the last

Ionic Size

- Cations are **smaller** than their neutral atom counterparts
  - Electrons are removed
    - Results in a decrease in electron repulsion
      - Allows nuclear charge to pull electrons closer
- Anions are **bigger** than their neutral atom counterparts
  - Electrons are added
    - Results in an increase in electron repulsion
      - Occupy more space
- Ionic size **increases** down a group
  - More energy levels
- Ionic size is slightly complicated across a period
  - Decreases among cations
  - Increases dramatically with the first anion (but decreases within anions)

Ionization Energy (IE)

- First ionization energy - the minimal energy needed to remove one of the outermost electrons from a neutral atom

    o Successive electron removal is called second ionization energy, third ionization energy, etc.

    o Successive ionization energies increase because each electron that is pulled away creates a larger positive charge (i.e. a higher effective nuclear charge)

- Ionization energies tend **increase** with atomic number within a period

    o More difficult to remove an electron that is closer to the nucleus

        ▪ Remember: There is a higher effective nuclear charge across a period

- Ionization energies tend to **decrease** down a group

Electron Affinity

This is not the same as electronegativity!

- More negative electron affinity value express that a more stable negative ion is formed

    o Negative values indicate that energy is released when the anion for that element forms

- General trend is that values become more negative from lower left to upper right

    o Highest electron affinities occur for F and Cl

Electronegativity

- Electronegativity is the measure of an atom's ability of to draw bonding electrons to itself in a molecule

- Electronegativity tends to increase from the lower-left corner to the upper-right corner of the periodic table

Metals, Non-Metals, and Metalloids

Metals

- Typical Properties
    - Shiny solids
    - High melting points
    - Good thermal & electrical conductors
    - Malleable
    - Lose their electrons to non-metals

Non-Metals

- Typical Properties
    - Not shiny
    - Low melting points
    - Poor thermal & electrical conductors
    - Crumbly solids or gases
    - Gain electrons from metals

Metalloids (Semi-metals)

- Elements that exhibits external characteristics of a metal, but behaves chemically as a nonmetal
- Typical Properties
    - Make good semiconductors
        - Have intermediate conductivity
    - Intermediate electronegativity values and ionization energies
    - Reactivity depends on the element with which they are reacting
    - Do not form multiple bonds
- Boiling points, melting points, and densities vary widely

Metallic Behavior Trends

- Metallic behavior decreases across (left to right) a period and increases down a group

CHAPTER 7 – CHEMICAL BONDING

Ionic Bonds

Ionic bonds are chemical bonds formed by the electrostatic attraction between positive and negative ions.

- Ionic bonds involve the transfer of electrons from one atom to another

    - Usually the transfer is from a metal from Group IA or IIA to a nonmetal from Group 7A or the top of Group 6A

- Number of electrons lost or gained by an atom is determined by its need to be isoelectronic with its nearest noble gas

    - Noble gas configurations are extremely stable

- Ionic bonds result in the formation of ions that are electrostatically attracted to one another

    - Anion – negatively charged ion

        - Size of an anion is larger than the original size of the neutral atom

    - Cation – positively charged ion

        - Size of a cation is smaller than the original size of the neutral atom

Properties of Ionic Compounds

- Hard: don't dent

- Rigid: don't bend

- Brittle: crack but don't deform

- The above properties are a result of the powerful attractive forces holding ions together

    - Moving ions out of position requires large amounts of energy to overcome the attractive forces

Covalent Bonds

- Two nonmetals often form covalent bonds

    - Share electrons since they have similar attractions for them

- Each nonmetal holds tightly to its own electrons
    - Shared electron pair spend most of their time between the two atoms
        - Electron pairs being shared are said to be localized
- How covalent bonds form
    - Distance between two nuclei decreases
        - Each starts to attract the other's electron(s)
        - Causes a decrease in potential energy
    - Atoms draw closer and closer together
        - Energy becomes progressively lower
    - Attractions increase but so does repulsions between electrons
    - At a particular internuclear distance, maximum attraction is achieved
        - Balance between nucleus-electron attractions
        - Balance between electron-electron and nucleus-nucleus repulsions
- Two sets of forces are involved within covalent compounds
    - Strong covalent bonding forces hold atoms together within the molecule
    - Weak intermolecular forces hold separate molecules near each other
        - Weak intermolecular forces between molecules are responsible for the observed physical properties of these molecules

Types of Covalent Bonds

- Coordinate covalent bond – covalent bond in which both of the shared electrons are donated by a single atom
- Double bond – sharing of two pairs of electrons between atoms
- Triple bonds – sharing of three pairs of electrons between atoms
- Polar covalent bond – covalent bond where the bonding electron spend more time closer to one of the atoms involved in the bonding
- Nonpolar covalent bond – covalent bond where the bonding electrons are shared equally

Bonding Pairs and Lone Pairs

- Shared electrons are considered as belonging entirely to each atom in a covalent bond

    o Shared electron pair simultaneously fills the outer level of both atoms

- An outer-level electron pair that is not involved in bonding is a "lone pair"

Metal-Metal Bonding

- Metals have low ionization energies

    o Lose electrons easily

    o Do not gain them readily

- Valence electrons are evenly distributed around metal-ion cores

    o Metal-ion cores consist of the nucleus plus the inner electrons

- Valence electrons are delocalized

    o Move freely throughout the metal

Predicting Ionic and Covalent Bonding

- Non-polar covalent bond

    o Typically electronegativity difference between the two atoms has to be less than 0.5 for non-polar bonds

- Polar covalent bonds

    o Electronegativity between the two atoms is different by a greater degree than 0.5 but less than 2.0

- Ionic bonds

    o Typically, difference in electronegativity is more than 2.0

Bond Length, Bond Order, and Bond Energy

- Bond order - measure of the number of bonding electron pairs between atoms

  o Single bonds have a bond order of 1

  o Double bonds have a bond order of 2

  o Triple bonds have a bond order of 3

  o Fractional bond orders are possible in molecules and ions that have resonance structures

- Bond length (aka bond distance) – distance between the nuclei in a bond

  o Bond length depends on bonds order

  o Increase in bond order means a shorter and stronger bond

- Bond Energy (BE) – measure of the amount of energy required to break apart one mole of covalently bonded gases

  o Quantity of heat absorbed to break reactant bonds is denoted $\Delta H^{\circ}_{reactants}$

  o Quantity of heat released to form product bonds is denoted $\Delta H^{\circ}_{products}$

  o Exothermic reactions

    ▪ ΔH°_{rxn} is negative

  o Endothermic reactions

    ▪ ΔH°_{rxn} is positive

  o $\Delta H^{\circ}_{rxn} = \sum BE_{reactant\ bonds\ broken} - \sum BE_{product\ bonds\ formed}$

CHAPTER 8 – GEOMETRY OF MOLECULES

Lewis Dot Structures

Lewis Structure of Carbon:

- Lewis dot structures represent electrons in the valence shell of an atom or ion as dots placed around the letter symbol of the element

  o Bonding electron pairs are represented by either two dots or a dash

Lewis Electron-dot Formula Example:

- Rules for Forming Lewis Structures

  o Calculate the number of valence electrons for the molecule

    ▪ Group # for each atom (1-8)

      • Gives valence electron number for each atom

      • Add all numbers up

    ▪ Add the charge of any anions

      • Example: an anion with a -2 charge has 2 extra electrons, you would add 2 to the total count

    ▪ Subtract the charge of any cations

      • Example: a cation with a +3 charge lacks 3 electrons, you would subtract 3 from the total count

  o Place the atom with the lowest group number and lowest electronegativity as the central atom

  o Arrange the other elements around the central atom

  o Distribute electrons to atoms surrounding the central atom to satisfy the octet rule for each atom

  o Distribute the remaining electrons as pairs to the central atom

- o If the central atom is deficient in electrons, complete the octet for it by forming double bonds or possibly a triple bond

Octet Rule

The octet rule states that the tendency of atoms in a molecule is to have eight electrons in their outer shell.

- There are exceptions to this rule where the central atom may have more than eight electrons

- Generally, a nonmetal in the third period or higher can accommodate as many as twelve electrons, if it is the central atom

  - o These elements have unfilled "d" subshells that can be used for bonding

Resonance (Delocalized Bonding)

- Structures of some molecules can be represented by more than one Lewis dot formula

  - o Individual Lewis structures are called contributing structures

  - o Individual contributing structures are connected by double-headed arrows (aka resonance arrows)

  - o Molecule or ion is a hybrid of the contributing structures and displays delocalized bonding

    - ▪ Delocalized bonding is where a bonding pair of electrons is spread over a number of atoms

- Some resonance structures contribute more to the overall structure than others

  - o Determining which structures are more contributing

    - ▪ Structures where all atoms have filled valence shells

    - ▪ Structures with the greater number of covalent bonds

    - ▪ Structures with less charges

      - • Formal charges can help discern which structure is most likely (discussed later in this section)

    - ▪ Structures that carry a negative charge on the more electronegative atom

Example of Resonance Structures:

- Curved arrow – symbol used to the redistribution of valence electrons

  - Always drawn as noted in the figure below

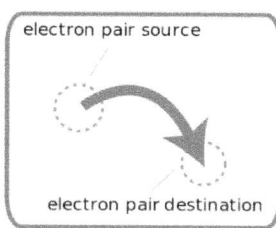

How Curved Arrows are Drawn:

Formal Charge

- An atom's formal charge is:

  - Total number of valence electrons

  - Minus all unshared electron

  - Minus ½ of its shared electrons

- Formal charges have to sum to the actual charge of the species

  - 0 charge for a molecule

  - Ionic charge for an ion

- Lewis structures with the smallest formal charge are the most likely to occur

Formal Charge vs. Oxidation Number

- Formal charges are used to examine resonance hybrid structures

  - Oxidation numbers are used to monitor redox reactions

- Formal charge

  - Bonding electrons are assigned equally to the atoms

    - Each atom has half the electrons making up the bond

- o Formal Charge = valence e⁻ − (unbonded e⁻ + ½ bonding e⁻)
- Oxidation Number
  - o Bonding electrons are transferred completely to the more electronegative atom
  - o Oxidation Number = valence e⁻ − (unbonded e⁻ + bonding e⁻)

Valence-Shell Electron Pair Repulsion Model (VSEPR)

VSEPR predicts the shapes of molecules and ions by assuming that the valence shell electron pairs are arranged as far from one another as possible.

- Molecular geometry – 3-D arrangement of atoms that constitute a molecule
  - o Shape of a molecule is determined by the positions of atomic nuclei relative to each other
- Rules to help discern electron pair arrangements
  - o Select the central atom
    - ▪ Place atom with the lowest group number in the center
      - • If atoms share the same group number place the atom with the higher period number in the center
  - o Draw the Lewis structure
  - o Determine the number of bonding electron pairs around the central atom
  - o Determine the number of non-bonding electron pairs
    - ▪ Multiple bonds are counted as a single electron pair
  - o Arrange the electron pairs as far apart as possible
    - ▪ Minimizes electron repulsions
  - o Add the number of bonding and lone pairs
  - o From that number and the number of lone pairs you can use the chart below to determine the geometry
    - ▪ For example: If you were given a molecule where the central atom had 2 bonding pairs and 1 non-bonding pair (total number =3), the molecular shape would be *bent/angular*

VSEPR Geometries

Steric No.	Basic Geometry 0 lone pair	1 lone pair	2 lone pairs	3 lone pairs	4 lone pairs
2	X—E—X 180° Linear				
3	120° Trigonal Planar	< 120° Bent or Angular			
4	109° Tetrahedral	< 109° Trigonal Pyramid	<< 109° Bent or Angular		
5	120°, 90° Trigonal Bipyramid	< 90°, < 120° Sawhorse or Seesaw	90° T-shape	180° Linear	
6	90° Octahedral	< 90° Square Pyramid	90° Square Planar	< 90° T-shape	180° Linear

Example Question: Determine the molecular shape of CO_2 according to VSEPR theory.

CO_2: O=C=O

- Central atom: C
- C has 2 bonding pairs, 0 non-bonding pairs
  - Remember that double bonds are counted as a single electron pair
- According to VSEPR model the bonds are arranged linearly
  - Bond angle = 180°
- Molecular shape is *linear*

Example Question: Determine the molecular shape of $COCl_2$ according to VSEPR theory.

Example of $COCl_2$:

$$\underset{\underset{\displaystyle Cl \qquad Cl}{|}}{\overset{\displaystyle O}{\underset{\displaystyle C}{\|}}}$$

- Central atom: C

- C has 3 bonding pairs, 0 non-bonding pairs

- According to VSEPR model the three groups of electron pairs are arranged in a trigonal plane

  - Bond angle = 120°

- Molecular shape is *trigonal planar*

Effect of Lone Pairs

- Lone pairs are less confined because they are held by a single nucleus

  - Allows them to exert a stronger repulsive force than a bonding pair

    - Results in a decrease in the angle between bonding pairs

Dipole Moment

- Dipole moment (μ) - measure of the degree of charge separation (molecular polarity) in a molecule

  - $\mu = Q*r$

    - Q = Charge

    - r = distance between the charges

- Polarity of individual bonds within a molecule can be viewed as vector quantities

  - Molecules that are perfectly symmetric have a zero dipole moment

    - These molecules are considered nonpolar

- o Molecules that exhibit any asymmetry in the arrangement of electron pairs will have some dipole moment

    - These molecules are considered polar

Net moment = ↑

Example of H_2O Polarity Vectors.

Molecular Geometry and Dipole Moment Relatedness

Molecular Geometry	Dipole Moment
Linear	Can Be nonzero
Linear	Zero
Trigonal Planar	Zero
Trigonal Planar Bent	Can Be nonzero
Tetrahedral	Zero
Tetrahedral Trigonal Pyramidal	Can Be nonzero
Tetrahedral Bent	Can Be nonzero
Trigonal Bipyramidal	Zero
Trigonal Bipyramidal SeeSaw	Can Be nonzero
Trigonal Bipyramidal T-Shaped	Can Be nonzero
Trigonal Bipyramidal Linear	Can Be nonzero
Octahedral	Zero
Octahedral Square Pyramidal	Can Be nonzero
Octahedral Square Planar	Zero

CHAPTER 9 – BONDING THEORIES

Valence Bond Theory

Valence bond theory is an attempt to explain the covalent bond from a quantum mechanical view.

- Orbitals (s, p, d, f) of the same type have the same energy

- According to the theory, a bond forms when two atomic orbitals overlap

  o Space formed by overlapping orbitals has a capacity for two electrons

    ▪ Must have opposite spins (+½ and -½)

  o Each orbital forming the bond has at least one unfilled slot to accommodate the electron being shared

  o Bond strength depends on the attraction of the nuclei for the shared electrons

    ▪ Greater orbital overlap = stronger bond

  o Overlap depends on the shapes and direction of the orbitals

Hybrid Orbitals

- Quantum mechanical calculations show that if specific combinations of orbitals are mixed, "new" atomic orbitals are formed

  o These new orbitals are called hybrid orbitals

- Types of hybrid orbitals

  o Each type has a unique geometric arrangement

Hybrid Orbitals (Hybridization)	Geometric Arrangements	Number of Hybrid Orbitals Formed by Central Atom	Example
sp	Linear	2	Be in BeF_2
sp^2	Trigonal planar	3	B in BF_3
sp^3	Tetrahedral	4	C in CH_4
sp^3d	Trigonal bipyramidal	5	P in PCl_5
sp^3d^2	Octahedral	6	S in SF_6

- Hybrid orbitals are used to describe bonding that is obtained by taking combinations of atomic orbitals of an isolated atom

- Steps for determining bonding description

    o Write the Lewis dot formula for the molecule

    o Then use the VSEPR theory to determine the arrangement of electron pairs around the central atom

    o From the geometric arrangement, determine the hybridization type

    o Assign valence electrons to the hybrid orbitals of the central atom one at a time

        ▪ Pair only when necessary

    o Form bonds to the central atom by overlapping singly occupied orbitals of other atoms with the singly occupied hybrid orbitals of the central atom

Multiple Bonds

- Orbitals can overlap two ways

    o Side to side

    o End to end

- Two types of covalent bonds

    o Sigma bonds (C-C)

        ▪ Formed from an overlap of one end of the orbital to the end of another orbital

    o pi bonds (C=C)

        ▪ Formed when orbitals overlap side to side

        ▪ Creates two regions of electron density

            • One above and one below

- Double bonds always consist of one sigma bond and one pi bond

Molecular Orbital Theory

- As atoms approach each other and their atomic orbitals overlap, molecular orbitals are formed

    o Only outer (valence) atomic orbitals interact enough to form molecular orbitals

- Combining atomic orbitals to form molecular orbitals involves adding or subtracting atomic wave functions

- Adding wave functions

    o Forms a bonding molecular orbital

    o Electron charge between nuclei is dispersed over a larger area than in atomic orbitals

    o Molecular orbitals have lower energy than atomic orbitals

        ▪ Reduction in electron repulsion

    o Bonding molecular orbital is more stable than atomic orbital

- Subtracting Wave Functions

    o Forms a antibonding molecular orbital

    o Electrons do not shield one nuclei from the other

        ▪ Results in increased nucleus-nucleus repulsion

    o Antibonding molecular orbitals have a higher energy than the corresponding atom orbitals

    o When the antibonding orbital is occupied, the molecule is less stable than when the orbital is not occupied

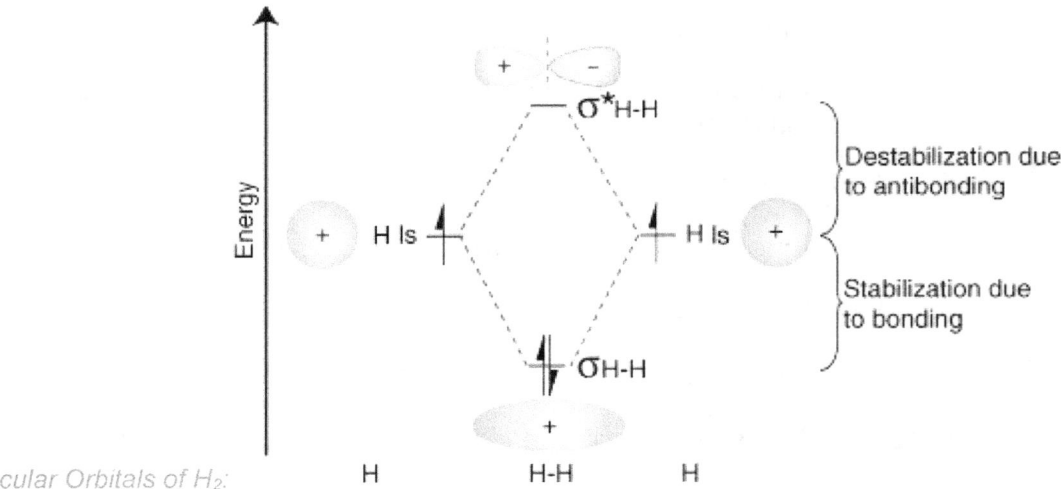

Molecular Orbitals of H₂:

CHAPTER 10 - GASES AND GAS LAWS

Properties of Gases

- Flow freely

- Relatively low densities

- Gases form homogenous mixtures with each other

- Volume of gases change significantly with pressure

- Volume of gases changes significantly with temperature

  - Under high temperatures gases expand

  - Under low temperatures gases contract

Pressure

- Force (F) - a function of the mass of an object under acceleration

  - F = Mass x Acceleration

- Pressure (P) - force exerted per unit area of surface by molecules in motion

  - $P = \frac{Force}{Area} = \frac{ma}{A} = \frac{mg}{A} = \frac{mg}{\frac{v}{h}} = \frac{mgh}{v} = dgh$

    - A = area

    - m = mass

    - a = acceleration

    - d = density

    - g = acceleration due to gravity = 9.81 m/s^2

    - h = height of column (m)

  - Pressure has many units

    - SI unit is Pascal (Pa) = 1 kg•m^{-1}•sec^{-2}

    - Atmosphere and torr are commonly used

      - 1 atmosphere (atm) = 760 mm Hg = 760 torr

Gas Laws

The behavior of gas can be described by pressure (P), temperature (T), volume (V), and molar amount (n). If you hold any of the two variables constant, it allows for determination of a relationship between the other two.

- Ideal gas – gas that exhibits linear relationships among pressure, temperature, volume, and molar amount
    - Ideal gases don't actually exist
        - But simple gases behave ideally under normal temperatures and pressures

- Molar gas volume (V_m) – volume of one mole of gas

- Volumes of gases are often compared at standard temperature and pressure (STP)
    - Standard Temperature = 0°C (273 K)
    - Standard Pressure = 1 atm
    - V_m at STP = 22.4 L/mol

- Boyles Law
    - Volume of a sample of gas at a constant temperature is inversely related to the applied pressure
    - $P_1V_1 = P_2V_2$ or $\dfrac{P_1}{P_2} = \dfrac{V_1}{V_2}$

- Charles Law
    - Volume of a sample of gas at constant pressure is directly proportional to the absolute temperature
    - $\dfrac{V_1}{T_1} = \dfrac{V_2}{T_2}$

- Avogadro's Law
    - Equal volumes of different gases at the same temperature and pressure contain equal number of particles
    - $\dfrac{V_1}{n_1} = \dfrac{V_2}{n_2}$

- Combined Gas Law
  - For when P, V, and T are changing
    - $\dfrac{P_1 V_1}{T_1} = \dfrac{P_2 V_2}{T_2}$
- Ideal Gas Law
  - PV = nRT
    - R is the universal gas constant
    - R = 0.082058 L•atm•mol^{-1}•K^{-1} = 8.3145 J•mol^{-1}•K^{-1}
  - As long as you know three of the variables you can manipulate the ideal gas law to solve for the fourth
- Molar Mass from Ideal Gas Law
  - $M = \dfrac{mRT}{PV}$
- Density
  - $d = \dfrac{P(MM)}{RT}$
- Law of Partial Pressures
  - $P_i = \dfrac{n_i}{n_{tot}} P_{tot} = x_i P_{tot}$
    - $x_i = n_i / n_{tot}$

Kinetic-Molecular Theory of Gases

- A model based on actions of individual atoms
  - Gases consist of particles in constant motion
  - Pressure derived from bombardment with container
  - Kinetic energy formula: $E_k = \frac{1}{2} mv^2$

- Postulates of Kinetic Theory

  - Volume of particles is negligible

  - Particles are in constant motion

  - No inherent attractive or repulsive forces

  - The average kinetic energy of a collection of particles is proportional to the temperature (K)

- Molecular Motion in Gases

  - Diffusion – transfer of gas through space or another gas over time

  - Effusion – transfer of a gas from a region of high pressure to a region of low pressure

- Grahams Law of Effusion

  - $$\frac{effusion\ Rate_A}{effusion\ Rate_B} = \frac{N_A}{N_B}\sqrt{\frac{MM_B}{MM_A}} = \sqrt{\frac{MM_B}{MM_A}}$$

  - $$\frac{effusion\ Time_A}{effusion\ Time_B} = \sqrt{\frac{MM_A}{MM_B}}$$

CHAPTER 11 - THERMOCHEMISTRY

Thermochemistry

- In chemical reactions whenever matter changes composition, its energy content changes as well

- In some reactions the energy contained in the reactants is higher than the energy contained in the products

    - The excess energy is released as heat

- In other reactions the energy contained in the reactants is lower than the energy contained in the products

    - In these reactions, energy (heat) must be added before the reaction can proceed

- Physical changes also involve a change in energy

- Thermodynamics - science of the relationship between heat and other forms of energy

- Thermochemistry - study of the quantity of heat absorbed or exuded by chemical reactions

Energy

Energy is the potential or capacity to do work. Energy is a property of matter and comes in many forms.

- Forms of energy

    - Radiant energy - electromagnetic radiation

    - Thermal energy - associated with random motion of a molecule or atom

    - Chemical energy - energy stored within the structural limits of a molecule or atom

- Concepts of Energy

    - Kinetic energy (E_k) – energy possessed by an object due to its motion

        - $E_k = \frac{1}{2} mv^2$

- - Potential energy (E_p) – energy stored in matter because of its position or location

    - $E_p = mgh$

  - Internal energy (E_i or U_i) – energy associated with the random disordered motion of molecules

    - $E_i = E_k + E_p$

- Units of Energy

  - SI unit of energy is the Joule (J) = $kg \cdot m^2/s^2$

  - Calorie (cal) - amount of heat required to raise the temperature of one gram of water by one degree Celsius

    - 1 cal = 4.181 J

- When reactants interact to form products and the products are allowed to return to the starting temperature, the Internal Energy (E) has changed

  - ΔE = change in energy

    - Δ is the symbol for change

      - Δ = final value – initial value

  - $\Delta E = E_{final} - E_{initial} = E_{products} - E_{reactants}$

- If energy is lost to the surroundings

  - $E_{final} < E_{initial}$

  - $\Delta E < 0$

- If energy is gained from the surroundings

  - $E_{final} > E_{initial}$

  - $\Delta E > 0$

Heat of Reaction

In chemical reactions, heat is either transferred from the "system" to its "surroundings," or vice versa.

- Thermodynamic system - quantity of matter or the space under thermodynamic study

- Surroundings - everything in the vicinity of the thermodynamic system that interacts with the system

- Heat (q) - energy that flows into or out of a system because of a difference in temperature between the system and its surroundings
    - Heat flows from a region of higher temperature to a region of lower temperature
        - Once temperatures equalize, heat flow stops
    - When heat is released from the system to the surrounding
        - $q < 0$
        - Reaction is called an **exothermic reaction**
    - When heat is absorbed from the surrounding by the system
        - $q > 0$
        - Reaction is called an **endothermic reaction**
- Heat of reaction - the value of "q" required to return a system to a given temperature when the reaction goes to completion

Work

Internal energy is specifically defined as the capacity of a system to do work.

- Work – energy transferred when an object is moved by a force
    - $w = {}^{-}P\Delta V$
    - $\Delta E = q_p + w$
    - $\Delta E = q_p + {}^{-}P\Delta V$
    - $q_p = \Delta E + P\Delta V$
        - q_p - heat absorbed from the surroundings by the system
        - ΔE - change in internal energy
        - ΔV - change in volume
        - P – pressure

Enthalpy and Enthalpy Change

- Enthalpy (H) - an extensive property of a substance that is used to obtain the heat absorbed or exuded in a chemical reaction

    o $H = E + PV$

    o Enthalpy is a state function

        ▪ Property of a system that depends only on its state at the moment and is independent of any history of the system

    o Enthalpy is representative of the heat energy tied up in chemical bonds

- Change in enthalpy (ΔH) - heat added or lost by the system, under constant pressure

    o $\Delta H = \Delta E + P\Delta V$

    o $\Delta H = q_p$

    o Change in enthalpy is also called the enthalpy of reaction

    o $\Delta H_{rxn} = H_{(products)} - H_{(reactants)}$

- If the system has higher enthalpy at the end of the reaction

    o It absorbed heat from the surroundings

    o It is an endothermic reaction

    o $H_{final} > H_{initial}$

    o ΔH is positive ($+\Delta H$)

- If the system has lower enthalpy at the end of the reaction

    o It exuded heat to the surroundings

    o It is an exothermic reaction

    o $H_{final} < H_{initial}$

    o ΔH is negative ($-\Delta H$)

Thermochemical Equation

Thermochemical equations are chemical reaction equations with the enthalpy of reaction (ΔH_{rxn}) written directly after the equation.

- Example of a thermochemical equation

    o $2 H_{2 (g)} + O_{2 (g)} \rightarrow 2 H_2O_{(l)}$ $\Delta H_{rxn} = -571.6$ kJ

        ▪ The negative value for ΔH_{rxn} is telling you that heat is lost to the surrounding

        ▪ Also that the equation is exothermic

- Rules for manipulating thermochemical equations

    o If the thermochemical equation is multiplied by some factor, the value of ΔH for the new equation is equal to the ΔH in the original equation multiplied by that factor

    o If the chemical equation is reversed, the sign of ΔH must be reversed

        ▪ Example, if you were to reverse the direction of the equation mentioned above you would get:

            • $2 H_2O_{(l)} \rightarrow 2 H_{2 (g)} + O_{2 (g)}$ $\Delta H_{rxn} = +571.6$ kJ

Measuring Heats of Reaction

- Heat capacity – amount of heat required to raise the temperature of an object or substance

    o Varies between substances

- Molar heat capacity (C) – amount of heat required to raise the temperature of **one mole** of a substance by **one degree Celsius**

    o $q = nC\Delta T$

        ▪ $\Delta T = T_{final} - T_{initial}$

- Specific heat capacity (S) – amount of heat required to raise the temperature of **one gram** of a substance by **one degree Celsius**

    o $q = mS\Delta T$

        ▪ $\Delta T = T_{final} - T_{initial}$

        ▪ Units for S: J/g•°C

        ▪ m = grams of a sample

- Hess's Law of Heat Summation

  - For a chemical equation that can be written as the sum of two or more steps, the enthalpy changes for the individual steps can be summed (added) up to determine the enthalpy change for the overall equation

  - For coupled reactions, the individual enthalpy changes can be summed up to determine the enthalpy change for the overall reaction

Hess's Law Example Question: What is the standard enthalpy of reaction for the reduction of iron (II) oxide by carbon monoxide? $FeO_{(s)} + CO_{(g)} \rightarrow Fe_{(s)} + CO_{2(g)}$

- Given Information:

  - Equation 1: $3\ Fe_2O_{3(s)} + CO_{(g)} \rightarrow 2\ Fe_3O_{4(s)} + CO_{2(g)}$ $\Delta H = -48.26$ kJ

  - Equation 2: $Fe_2O_{3(s)} + 3\ CO_{(g)} \rightarrow 2\ Fe_{(s)} + 3\ CO_{2(g)}$ $\Delta H = -23.44$ kJ

  - Equation 3: $Fe_3O_{4(s)} + CO_{(g)} \rightarrow 3\ FeO_{(s)} + CO_{2(g)}$ $\Delta H = +21.79$ kJ

- Changes have to made to the above equations to equal the equation in the question

  - Reverse equation 3 and multiply it by two

    - Puts FeO on the reactant side and moves 2 Fe_3O_4 to the products

  - Reverse equation 1

    - Puts Fe_3O_4 on opposite side to cancel with the reverse of equation 3

  - Multiply equation 2 by three

    - Gives 3 Fe_2O_3 on the reactants that will be used to cancel

- New equations after changes

  - Equation 1: $2\ Fe_3O_{4(s)} + CO_{2(g)} \rightarrow 3\ Fe_2O_{3(s)} + CO_{(g)}$ $\Delta H = +48.26$ kJ

  - Equation 2: $3\ Fe_2O_{3(s)} + 9\ CO_{(g)} \rightarrow 6\ Fe_{(s)} + 9\ CO_{2(g)}$ $\Delta H = -70.32$ kJ

  - Equation 3: $6\ FeO_{(s)} + 2\ CO_{2(g)} \rightarrow 2\ Fe_3O_{4(s)} + 2\ CO_{(g)}$ $\Delta H = -43.58$ kJ

- Summing the three equations gives

  - $6\ FeO_{(s)} + 6\ CO_{(g)} \rightarrow 6\ Fe_{(s)} + 6\ CO_{2(g)}$ $\Delta H = -65.64$ kJ

- Dividing by six gives the equation in the question and the final answer

  - $FeO_{(s)} + CO_{(g)} \rightarrow Fe_{(s)} + CO_{2(g)}$ $\Delta H = -10.94$ kJ

Standard Enthalpies of Formation

- Standard state refers to the standard thermodynamic conditions

    o Pressure - 1 atm (760 mm Hg)

    o Temperature - 25°C (298 K)

- Enthalpy change for a reaction where reactants are in their standard states is called the "Standard Heat of Reaction"

    o $\Delta H°_{rxn}$

- Standard enthalpy of formation of a substance - enthalpy change for the formation of one mole of a substance in its standard state from its component elements in their standard states

    o Standard enthalpy of formation for a "pure" element (C, Fe, O, N, etc.) in its standard state is zero

- Law of Summation of Heats of Formation

    o The standard heat of reaction ($\Delta H°_{rxn}$) is equal to the total formation energy of the products minus the total formation energy of the reactants

        ▪ $\Delta H°_{rxn} = \sum n\Delta H°_f (products) - \sum m\Delta H°_f (reactants)$

        ▪ m and n are coefficients of the substances in the chemical equation

Gibbs Free Energy

Gibbs free energy can be used to determine the direction of the chemical reaction under given conditions.

- $\Delta G = \Delta H - T\Delta S$ or $\Delta G = G_{products} - G_{reactants}$

    o G = Gibbs free energy (J/mol)

    o H = enthalpy (J/mol) - total energy content of a system

    o S = entropy (J/K*mol) - measure of disorder or randomness (how energy is dispersed)

    o T = Temperature (K)

        ▪ As T increases so does S

- +ΔG means energy must be put into the system
  - Indicates that a process is **nonspontaneous or endergonic**
  - Indicates that the position of the equilibrium for a reaction favors the products
- -ΔG means energy is released by the system
  - Indicates that a process is **spontaneous or exergonic**
  - Indicates that the position of the equilibrium for a reaction favors the reactants
- ΔG = 0 indicates that the system is at equilibrium

Important: ΔG only indicates if a process occurs spontaneously or not, but does **not** indicate anything about how fast a process occurs.

CHAPTER 12: SOLUTIONS

Solutions

- Solutions are composed of a solute and a solvent

    o Solutions are homogeneous mixtures

    o Solutes are minor components in a solution (present in smaller amounts)

    o Solvents are substances in which a solute is dissolved

- Solution composition is expressed by:

    o Mass percent

        ▪ $mass\ \% = \dfrac{mass\ solute}{mass\ of\ solution} \times 100\%$

        ▪ $mass\ \% = \dfrac{mass\ solute}{mass\ solvent\ +\ mass\ solute} \times 100\%$

    o Mole fraction

        ▪ $X_{solute} = \dfrac{mass\ solute}{mass\ solvent\ +\ mass\ solute}$

        ▪ $X_{solute} = n_{solute} / n_{total}$

    o Molarity

        ▪ $M = \dfrac{moles\ solute}{Liters\ of\ solution} = \dfrac{mol}{L}$

    o Molality

        ▪ $m = \dfrac{moles\ solute}{kg\ solvent} = \dfrac{mol}{kg}$

Conversion between Molarity and Molality

Density must be known to convert from molarity to molality directly.

Example Question: What is molality of 2.00 M NaCl$_{(aq)}$ solution with a density of 1.08 g/mL?

- Determine the mass of 2.00 mol of NaCl

    o 2.00 mol x (58.5 g /mol) = 117 g NaCl

- Assume you have 1.000L (1000 mL) of the 2.00 M NaCl solution
  - You can assume any amount if it is not explicitly stated in the question
  - Hint: it is best to assume an amount that is easy to work with
    - To convert molarity to molality assume 1.000 L of solution
    - To convert molality to molarity assume 1.000 kg of solvent
- Use the density given in the problem to determine the total mass of the solution
  - 1000 mL x (1.08 g/mL) = 1080 g total mass
- Determine the mass of the solvent
  - Earlier we calculated that we had 117 g NaCl and the total mass of the solution was 1080 g
  - So, to determine the mass of the solvent we will simply subtract the difference
    - 1080 g of solution – 117 g NaCl = 963 g solvent (0.963 kg solvent)
- Finally, use the formula for molality to determine the answer
  - $m = \dfrac{moles\ solute}{kg\ solvent} = \dfrac{2.00\ mol}{0.963\ kg} = 2.08\dfrac{mol}{kg}$

Concentrated vs. Dilute Solutions

- Concentrated solution – a solution that contains a high amount of solute
  - More solute particles per unit volume
- Dilute solution – a solution that contains a low amount of solute
  - Fewer solute particles per unit volume
- Saturated solution - a solution that contains the maximum amount of solute that can be dissolved by the solvent

Dilutions

- Extra solvent is added to a solution to dilute it
  - The amount of solute in the solution remains the same

- Use the following formula to solve dilution problems:
  - $M_1V_1 = M_2V_2$
    - M_1 – is the concentration of stock solution
    - V_1 – is the volume of stock solution
    - M_2 – is the concentration of the final solution
    - V_2 – is the volume of the final solution

Henry's Law

*Formulated by William **Henry** in 1803, it states: "At a constant temperature, the amount of a given gas that dissolves in a given type and volume of liquid is directly proportional to the partial pressure of that gas in equilibrium with that liquid."*

- $C = k_x P_{gas}$
  - C – solubility of a gas at a fixed temperature in a particular solvent
  - k - Henry's Law constant
  - P_{gas} – partial pressure of the gas

Colligative Properties

Colligative properties refer to properties of solutions that depend upon the ratio of the number of solute particles to the number of solvent molecules in a solution. These properties don't depend on the type of chemical species present.

- Common colligative properties
  - Vapor pressure lowering
  - Freezing point depression
  - Boiling point elevation
  - Osmotic pressure (discussed in detail in its own section)

Vapor Pressure Lowering

- Raoult's Law says that if you add a nonvolatile solute to a solvent, you will cause the vapor pressure of the solute to be lower

- Raoult's Law equation:

  ○ $P_{solution} = X_{solvent} P^0_{solvent}$

    ▪ $P_{solution}$ - vapor pressure of solution

    ▪ $X_{solvent}$ - mole fraction of solvent in solution

    ▪ P^0 solvent - vapor pressure of pure solvent

  ○ Because $X_{solvent}$ is a mole fraction (a number between 0 and 1), $P_{solution}$ is always lower than $P^0_{solvent}$

Freezing Point Depression

- Property based on the observation that freezing points of solutions are all lower than that of the pure solvent and is directly proportional to the molality of the solute

- Formula: $\Delta T_f = T_{f(solvent)} - T_{f(solution)} = K_f \times m$

  ○ ΔT_f - freezing point depression

  ○ $T_{f(solvent)}$ – freezing point of the solvent

  ○ $T_{f(solution)}$ – freezing point of the solution

  ○ K_f = freezing point depression constant

  ○ m – molality

Boiling Point Elevation

- Property based on the observation that the boiling point of a solvent is higher when another compound is added (i.e. solution has a higher boiling point than a pure solvent)

- Formula: $\Delta T_b = K_b \times m$

  ○ ΔT_b – boiling point elevation

  ○ K_b = boiling point elevation constant

  ○ m – molality

Osmotic Pressure

Semipermeable membranes stop solute molecules or ions from passing through but allow passage of solvent molecules. Solvent molecules such as water will go through membranes to dilute a solution unless a pressure equal to the osmotic pressure is applied to stop the flow. So, osmotic pressure is defined as the minimum pressure which needs to be applied to a solution to prevent the inward flow of water across a semipermeable membrane.

- Hypertonic - refers to a solution that has higher osmotic pressure than a particular fluid (e.g. intracellular fluid)

- Isotonic - refers to a solution that has the same osmotic pressure than a particular fluid (e.g. intracellular fluid)

- Hypotonic - refers to a solution that has the lower osmotic pressure than a particular fluid (e.g. intracellular fluid)

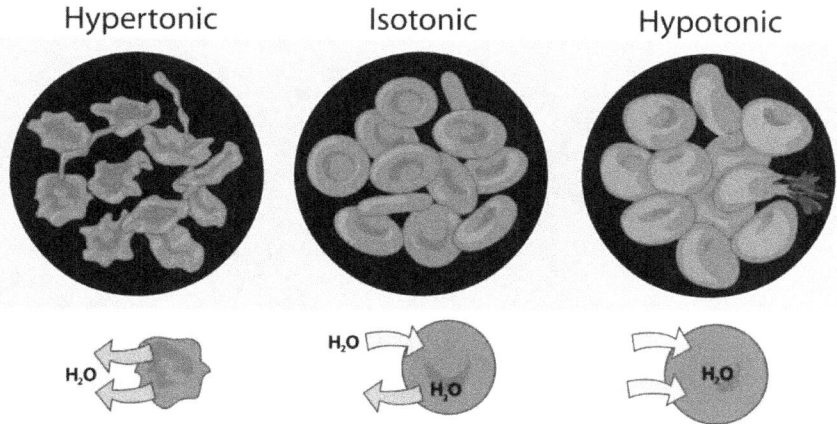

- Osmotic pressure (π) formula

    - $\pi V = nRT$ or $\pi = MRT$

        - n = moles of solute

        - V = volume of solution (L)

        - R = gas constant (0.08206 L·atm/mol·K)

        - T = temperature in Kelvin

        - M = molarity

Hydrophobic Effect and Amphiphilic Molecules

The hydrophobic effect is the tendency of nonpolar substances to aggregate in aqueous solution and exclude water molecules. Amphiphilic molecules have both hydrophobic and hydrophilic parts.

- Hydrophobic – water "hating"

- In contrast, hydrophilic – water "loving"

Amphiphilic Molecules in Water

- Nonpolar tails (hydrophobic portion) point away from water

- Polar heads (hydrophilic portion) are exposed to water

- Different amphipathic molecules aggregate in different ways based on the number of tails and size of the polar head group

- Micelles form when there is only 1 tail and take spherical form in aqueous solutions

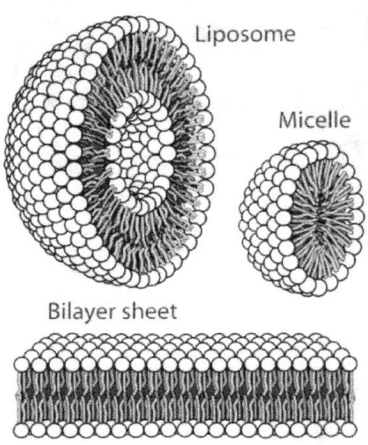

Liposome

Micelle

Bilayer sheet

CHAPTER 13 – CHEMICAL KINETICS

Introduction

- Reactions occur at different rates

    - Some are very quick, some are very slow, and many fall somewhere in between

- Knowing the rate of a reaction helps chemists plan out experiments and plan reactions accordingly

    - If you understand what contributes to rate, you can exert some control over a reaction

- Chemical equations (e.g., $Al_2O_3 \rightarrow Al + O_2$) don't tell you anything about how fast the reaction occurs

    - Some reactions occur in a series of smaller steps that lead to the final product

Reaction Rates

Generally, rates are defined as the change of something divided by change in time. This is true of reaction rates as well.

- Rate of a reaction can be written with respect to any compound in that reaction

    - But, there can only be one numerical value for a rate of reaction

- If you plot average rate data as a function of time, you will see that the reaction rate constantly changes (consider the graph below)

    - As you might notice, rate depends on the concentration of the reactants

- Since the rate of a reaction is effected by the concentration of the reactants we can write mathematical relationships linking the concentration of reactants with how fast the reaction occurs (i.e. we can write rate laws)

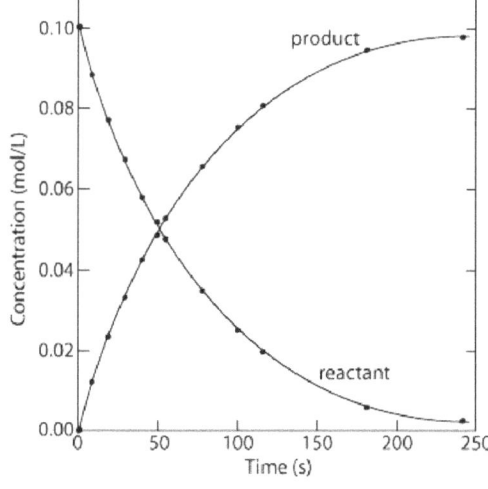

- General Reaction Rates
    - Consider the general chemical equation: aA + bB → cC + dD
    - Rate for this reaction would be defined as:
        - $$Rate = -\frac{1}{a}\left(\frac{\Delta[A]}{\Delta t}\right) = -\frac{1}{b}\left(\frac{\Delta[B]}{\Delta t}\right) = \frac{1}{c}\left(\frac{\Delta[C]}{\Delta t}\right) = -\frac{1}{d}\left(\frac{\Delta[D]}{\Delta t}\right)$$
- A simple rate law example:
    - Consider the decomposition reaction where: A → products
    - If the reverse reaction is negligible, then the rate law is: Rate = k[A]^n
        - k is called the **rate constant**
        - n is called the **reaction order**

Reaction Orders

- Reaction order (denoted as "n") determines how the rate depends on the concentration of the reactant
    - n = 0, zero order, rate is independent of [A]
    - n = 1, first order, rate is directly proportional to [A]
    - n = 2, second order, rate is proportional to the square of [A]
- Each order results in a different type of curve when graphed

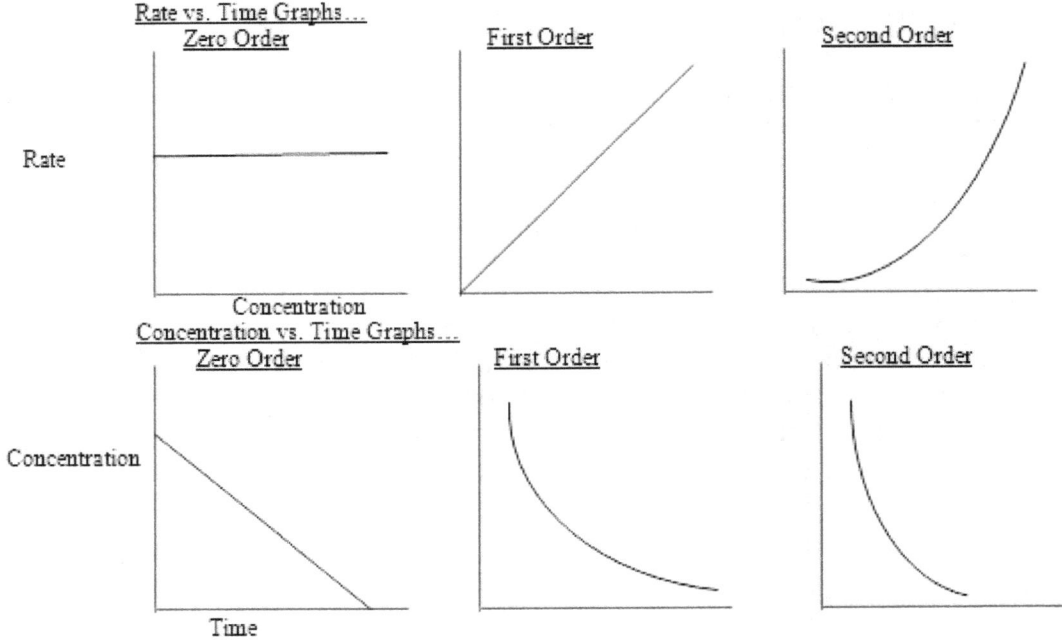

- ***Important***: You can only determine reaction orders through an experiment
  - Reaction orders are **not** related to the stoichiometry of a reaction

Steps for Finding Rate Law

- Pick two solutions where one reactant stays same, but another changes

- Write the rate law for both using as much information as you have

- Form a ratio from the two and solve for an order

- Repeat the 3 steps above for another pair of solutions

- Use any reaction to get the value of k

An Example Problem

- Imagine we are considering the general reaction: A + B → products

- And that we determined the following information from an experiment:

Exp.	Initial A (mmol/L)	Initial B (mmol/L)	Init. Rate of Formation of products (mM min^{-1})
1	4.0	6.0	1.60
2	2.0	6.0	0.80
3	4.0	3.0	0.40

- Look at experiments 1 and 2

  - From experiment 2 to 1, we see that the concentration of A doubles (while B is held constant) and the rate also doubles

  - Doubling of the rate with a doubling of the concentration shows that the reaction is first order with respect to A

- Next look at experiments 1 and 3

  - Concentration of B is halved (while A is held constant)

  - When B is halved, the overall rate drops by a factor of 4 (which is the square of 2)

  - This shows the reaction is second order with respect to B

- The rate law would be written as: rate = k [A] [B]^2

- You can use any reaction to get the value of k (I will use experiment 1)

  - Rate = k [A] [B]^2

  - 1.60 mM min^{-1} = k (4.0 mM) (6.0 mM)^2

  - k = 0.011 mM^{-2} min^{-2}

Integrated Rate Law and Half Life Formulas

Order	Integrated Rate Law	½ Life
Zero	$[A]_t = kt + [A]_o$	$\dfrac{[A]_o}{2k} = t_{1/2}$
First	$\ln([A]_t) = -kt + \ln([A]_o)$	$t_{1/2} = \dfrac{0.693}{k}$
Second	$\dfrac{1}{[A]_t} = kt + \dfrac{1}{[A]_o}$	$\dfrac{1}{k[A]_o} = t_{1/2}$

- Note that the integrated rate law equations are in the form y = mx + b

  - y = mx + b is the formula for a straight line

  - So, the plot of ln[A] vs. time for the reactions will yield a straight line

- For 2nd order, half-life depends on initial concentration

  - As concentration decreases, the half-life increases

- Half-life for zero order reactions depends on concentration as well

- However, notice that the half-life doesn't depend on reactant concentration for the 1st order reactions

  - Half-life for a 1st order reaction is constant

Temperature and Rate

- Generally, rates of reaction are sensitive to temperature

  - Rate = k[A]^n, so where do we factor in temperature?

    - It is reflected in the constant k

    - Generally, increasing temperature increases k

Chemistry of Catalysis

Catalysts do not change the direction of a chemical reaction and they have no effect on equilibrium!

- Function by lowering the activation energy, which speeds up the reaction

- As the reaction progresses; reactants become products

  - Depicted as a reaction coordinate diagram

  - Progress of the reaction is indicated on the x-axis

  - Free energy (G) is indicated on the y-axis

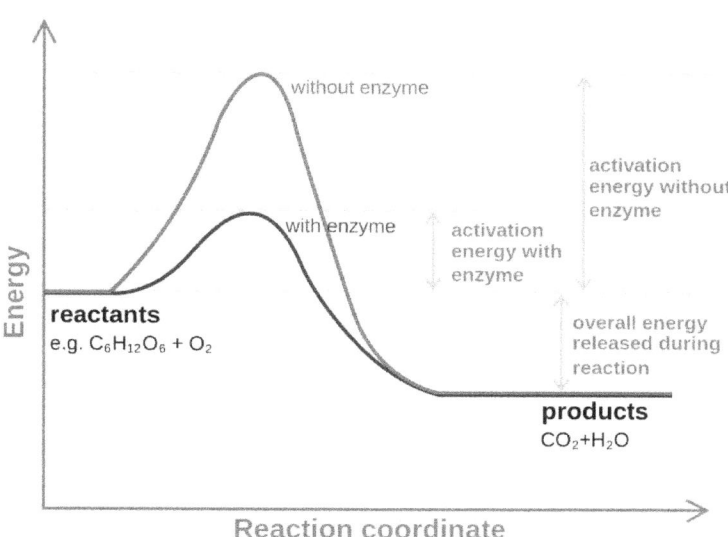

- Reactants pass through the transition state (‡) and become products

  - Enzymes increase the reaction rate by binding tightly to the transition state and stabilizing it

Spontaneous vs. Non-spontaneous Reactions

- Spontaneous if ΔG_{rxn} negative

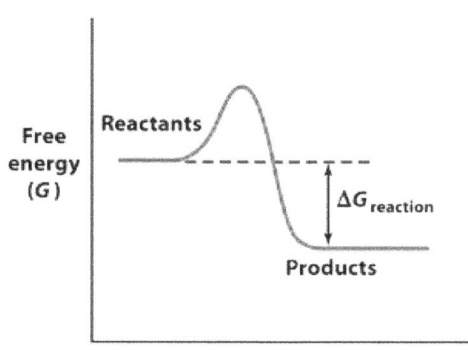

General Spontaneous Reaction Diagram:

- Non-spontaneous if ΔG_{rxn} positive

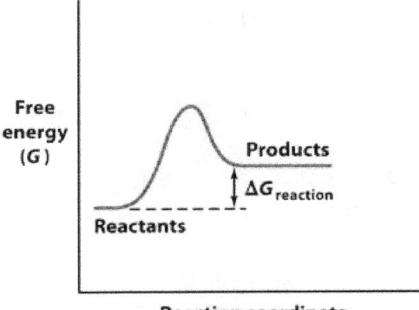

General Non-spontaneous Reaction Diagram:

- Enzymes (catalysts) lower the energy barrier

  o Make it easier to reach the transition state

CHAPTER 14 – CHEMICAL EQUILIBRIUM

What is Equilibrium?

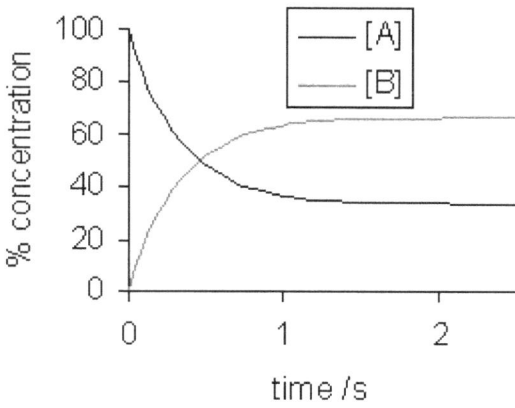

- As a system is approaching equilibrium, both the forward and reverse reactions are occurring at different rates

- Chemical equilibrium is established when a reaction and its reverse reaction occur at the same rate

- Once equilibrium is established, the amount of both the reactant and product remains constant

- In a system at equilibrium, both the forward and reverse reactions are running simultaneously so we write the chemical equation for the reaction with a double arrow

  - Example: $H_2 \rightleftharpoons 2H$

Equilibrium Constant

- Consider the reaction: $N_2O_4(g) \rightleftharpoons 2NO_2(g)$

  - Rate law for the forward reaction would be: rate = k_f [N_2O_4]

  - Rate law for the reverse reaction would be: rate = k_r [NO_2]^2

- o At equilibrium the two rates would be the same so we can rearrange the equations to get:

$$K_{eq} = \frac{k_f}{k_r} = \frac{[NO_2]^2}{[N_2O_4]}$$

  - ■ K_{eq} (aka K_c) is the equilibrium constant

- In general, the reaction: $aA + bB \rightleftharpoons cC + dD$

  - o Results in the equilibrium expression:
$$K_c = \frac{[C]^c [D]^d}{[A]^a [B]^b}$$

- Equilibrium can be reached from either the forward or reverse direction

  - o K_c, the final ratio of $[NO_2]^2$ to $[N_2O_4]$, will reach a constant no matter what the initial concentrations of NO_2 and N_2O_4 are (as long as time is held constant between them)

  - o Also note that the equilibrium constant of a reaction in the reverse reaction is the reciprocal of the equilibrium constant of the forward reaction

- If $K \gg 1$, the reaction is said to be *product-favored*

  - o Products predominate at equilibrium

- If $K \ll 1$, the reaction is said to be *reactant-favored*

  - o Reactants predominate at equilibrium

- If a reaction consists of many individual steps, you can add the equilibrium constants for the individual steps to determine the equilibrium constant for the entire reaction

- Pressure is proportional to concentration for gases, because of this, the equilibrium expression can also be written in terms of partial pressures (instead of concentration)

  - o
$$K_p = \frac{(P_C)^c(P_D)^d}{(P_A)^a(P_B)^b}$$

  - o K_p and K_c are related to one another by the equation: $K_p = K_c \, (RT)^{\Delta n}$

    - ■ Δn = *(moles of gaseous product) – (moles of gaseous reactant)*

Homogeneous and Heterogeneous Equilibrium

A homogeneous equilibrium is an equilibrium where all reagents and products are found in the same phase (solid, liquid, or gas). A heterogeneous equilibrium is an equilibrium where they are in different phases.

- Concentrations of liquids and solids can be obtained by the following:
$$\frac{density}{molar\ mass} = \frac{g/L}{g/mol} = \frac{mol}{L}$$

- Concentration of solids and liquids are not used to form an equilibrium expression

  - Consider the reaction: $PbCl_{2\ (s)} \rightleftharpoons Pb^{2+}{}_{(aq)} + 2\ Cl^{-}{}_{(aq)}$

  - The equilibrium constant for the reaction would be:

    - $K_c = [Pb^{2+}][Cl^{-}]^2$

Reaction Quotient (Q)

- To calculate Q, you have to substitute the initial concentrations of reactants and products into the equilibrium expression

- Q gives the same ratio as the equilibrium expression but for a system that is **not** at equilibrium

  - If Q = K, the system is at equilibrium

  - If Q > K, there is more product, and the equilibrium shifts to the reactants

  - If Q < K, there are more reactants, and the equilibrium shifts to the products

  - The shifting of equilibrium is Le Châtelier's Principle

    - Essentially the principle states that equilibrium position shifts to counteract the effect of a disturbance (change in temperature, concentration, etc.)

CHAPTER 15 – ACID BASE EQUILIBRIUM

Definitions and Conventions

Acid-base reactions are a type of chemical process typified by the exchange of one or more hydrogen ions (i.e. exchange/transfer of a proton).

- Arrhenius definition

    o Acid – substance that produces H^+ ions in aqueous solution

    - We now know that H^+ reacts immediately with a water molecule to produce a hydronium ion (H_3O^+)

    o Base – substance that produces OH^- ions in aqueous solution

- Bronsted-Lowry definition

    o Acid – proton donor

    o Base – proton acceptor

    o Bronsted-Lowry definition does not require water as a reactant

- Conjugate acids and bases

    o Conjugate base – species that is formed when an acid donates a proton to a base

    o Conjugate acid – species that is formed when a base accepts a proton from an acid

    o Conjugate acid-base pair – pair of molecules or ions that can be interconverted through the transfer of a proton

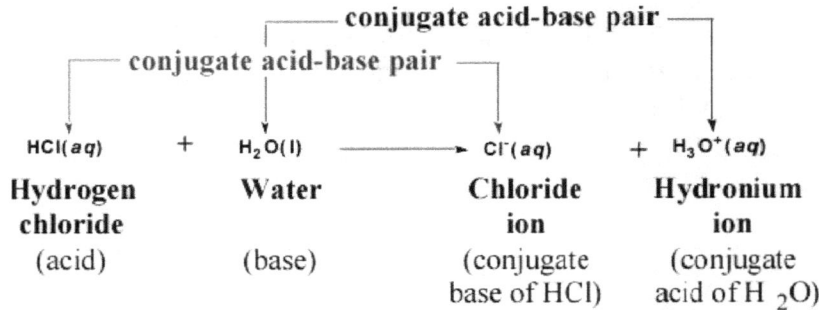

- Curved arrows are used to show the flow of electrons in an acid-base reaction

| Acetic acid | Ammonia | Acetate ion | Ammonium |
| (proton donor) | (proton acceptor) | | ion |

- Neutralization is the reaction of an H^+ (H_3O^+) ion from the acid and the OH^- ion from the base to form water, H_2O

    o Neutralization reaction is exothermic and releases approximately 56 kJ per mole of acid and base

- Determining acidic, basic, and neutral from concentration of H_3O^+ and OH^-

    o Neutral: $[H_3O^+] = [OH^-]$

    o Acidic: $[H_3O^+] > [OH^-]$

    o Basic: $[H_3O^+] < [OH^-]$

Strengths of Acids and Bases

- Strength of an acid is expressed by an equilibrium constant

    o Equilibrium expression for the dissociation of an uncharged acid (HA)

$$HA + H_2O \rightleftharpoons A^- + H_3O^+$$

$$K_{eq} = \frac{[H_3O^+][A^-]}{[HA][H_2O]}$$

    o K_a, the acid dissociation constant, is given by:

$$K_a = K_{eq}[H_2O] = \frac{[H_3O^+][A^-]}{[HA]}$$

        ▪ This is because the concentration of water is high, and does not significantly change during the reaction, so its value is absorbed into the constant

        ▪ The stronger the acid, the larger the K_a, and the more it will dissociate in solution

- Strong acids completely dissociate into ions in water

  - Strong acids are HI, HBr, $HClO_4$, HCl, $HClO_3$, H_2SO_4, and HNO_3

    - Their conjugate bases are weak

- Weak acids only partially dissociate into ions in water

- Polyprotic acids are acids that are capable of losing more than a single proton per molecule during an acid-base reaction

  - Phosphoric acid is a weak acid that normally only loses one proton but it will lose all three when reacted with a strong base at high temperatures

  - If the difference between the K_a for the first dissociation and subsequent K_a values is 10^3 or more, the pH generally depends **only** on the first dissociation

- In any acid-base reaction, the equilibrium favors the reaction that moves the proton to the stronger base

- The more polar the H-X bond and/or the weaker the H-X bond, the more acidic the compound

- Strong base – a base that is present almost entirely as ions (one of the ions is OH^-)

  - Strong bases are NaOH, KOH, LiOH, RbOH, CsOH, $Ca(OH)_2$, $Ba(OH)_2$, and $Sr(OH)_2$

- Weak base – a base that only partially ionizes in water

  - The general weak base reaction is written as:
    $$\ddot{B} + H_2O \rightleftharpoons HB^+ + OH^-$$

  - The equilibrium constant expression for this reaction is:
    $$K_c = K_b = \frac{[HB^+][OH^-]}{[B]}$$

    - K_b is called the base-dissociation constant

- K_a and K_b can be related to one another using the following formula:

  - $K_a \times K_b = K_w$

  - K_w is the ionization constant for water at 25 °C

    - K for water is: $K = \frac{[H^+][OH^-]}{[H_2O]}$ or $K_w = [H^+][OH^-] = 1.0 \times 10^{-14}$

Finding Concentration of Species in Solution from K_a

- Given: 0.10 M HNO_2 (nitrous acid), $K_a = 4.5 \times 10^{-4}$

- Set up a table to help you keep track of what is happening during the reaction

	$HNO_2 \rightleftharpoons$	H^+ +	NO_2^-
Initial Concentration	0.10 M	0.00 M	0.00 M
Change in Concentration	-x	+x	+x
Equilibrium Amount	0.10 M - x	x	x

  - Some of the reactant will become product, that is why the change in concentration is negative x

  - We are forming some amount of product in this reaction so the change in concentration is positive x

- We now need to solve for x, first set up the K_a equation

  - $K_a = \dfrac{[H^+][NO_2^-]}{[HNO_2]} = \dfrac{[x][x]}{[0.10-x]}$

  - $4.5 \times 10^{-4} = \dfrac{x^2}{0.10-x}$

  - The simple approach:

    - It is accepted that as long as X< 5% of [HA], where [HA] = concentration of the acid, you can assume x is negligible and that (0.10 - x) = 0.10, making the K_a equation:

    - $4.5 \times 10^{-4} = \dfrac{x^2}{0.10}$

    - $4.5 \times 10^{-5} = x^2$

    - $6.7 \times 10^{-3} = x$

  - The exact approach (quadratic formula):

    - However, if you can't assume x is going to be smaller than 5% you have to set up the exact formula

    - The quadratic equation is: $ax^2 + bx + c = 0$

    - $4.5 \times 10^{-4} = \dfrac{x^2}{0.10-x}$ $\rightarrow$ $x^2 + 4.5 \times 10^{-4}x - 4.5 \times 10^{-5} = 0$

      - a = 1, b = 4.5 x 10^{-4}, and c = -4.5 x 10^{-5}

- The quadratic formula:
$$x = \frac{-b \pm \sqrt{b^2 - 4ac}}{2a}$$

    - To solve, substitute in the values for a, b, and c

    - $x = \dfrac{-4.5 \; x \; 10^{-4} \pm \sqrt{(4.5 \; x \; 10^{-4})^2 - 4(1)(-4.5 \; x \; 10^{-5})}}{2(1)}$

    - $x = 6.5 \; x \; 10^{-3}$

        - Always use only the positive root, the negative root does not make sense in the context of these sort of problems

    - If you compare the answer from both approaches ($6.7 \; x \; 10^{-3}$ vs. $6.5 \; x \; 10^{-3}$) you can see that the answers are pretty much the same

        - However, remember that you can only use the simple approach if X< 5% of [HA]

- Since we found x, we only need to substitute the value into the "Equilibrium Amount" section of the table we set up earlier to find the concentration of species in solution

    - $[HNO_2] = (0.10 \; M - x) = (0.10 \; M - 6.5 \; x \; 10^{-3}) = 0.0935 \; M$

    - $[H^+] = x = 6.5 \; x 10^{-3} \; M$

    - $[NO_2^-] = x = 6.5 \; x 10^{-3} \; M$

pH and pOH

- pH is defined as the negative, base-10 logarithm of the hydronium ion concentration

    - $pH = -log[H_3O^+] \rightarrow [H_3O^+] = 10^{-pH}$

- In pure water:

    - $K_w = [H_3O^+][OH^-] = 1.0 \; x \; 10^{-14}$

    - Because in pure water $[H_3O^+] = [OH^-]$;

        - $[H_3O^+] = (1.0 \times 10^{-14})^{1/2} = 1.0 \; x \; 10^{-7}$

        - $pH = -log[1.0 \; x \; 10^{-7}] = 7.00$

        - 7.00 is considered neutral pH

- An acid has a higher $[H_3O^+]$ than pure water

  - pH is < 7

- A base has a lower $[H_3O^+]$ than pure water

  - pH is > 7

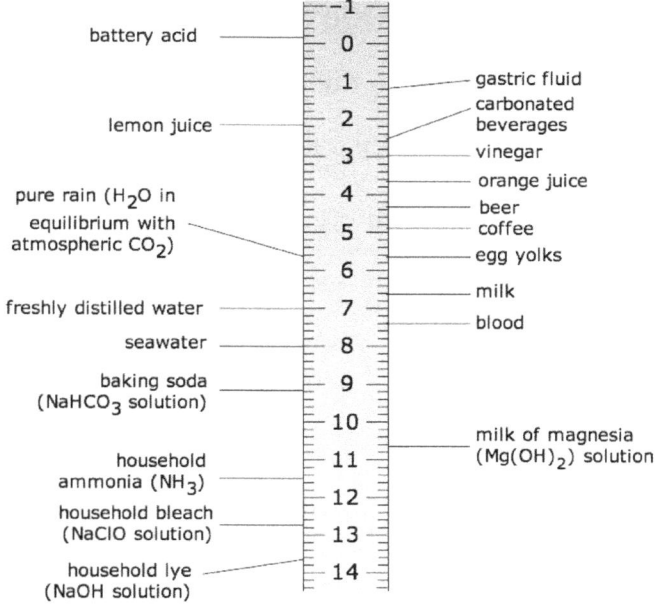

- p in pH is a clue to take the negative log of the quantity, this is true for pOH and pK_w:

  - $pOH = -\log[OH^-] \rightarrow [OH^-] = 10^{-pOH}$

  - $pK_w = -\log(K_w) \rightarrow K_w = 10^{-pK_w}$

pK_a and Trends

- $K_a = 10^{-pka}$

- $pK_a = -\log(K_a)$

  - Lower the pK_a, the stronger the acid

    - Higher the pK_a, the weaker the acid

  - Lower the pK_a, the weaker the conjugate base

    - Higher the pK_a, the stronger the conjugate base

- Equilibrium favors the side of the weakest acid and weakest base

  - Equilibrium favors the side with the highest pK_a

    - Thus, pK_a can be used to predict in which direction equilibrium lies

Percent Ionization Formulas

- $Percent\ ionization = \dfrac{amount\ ionized}{total\ in\ solution} \times 100\%$

- $Percent\ ionization = \dfrac{[A^-]}{[HA]+[A^-]} \times 100\%$

Reactions of Anions and Cations with Water

- Anions are bases

    - They can react with water in a hydrolysis reaction to form OH^- and the conjugate acid:

$$X^-(aq) + H_2O(l) \rightleftharpoons HX(aq) + OH^-(aq)$$

- Cations with acidic protons (like NH_4^+) lower the pH of a solution because they release H^+ ions in solution

- Most metal cations that are hydrated in solution also lower the pH of the solution

    - Act by associating with H_2O and making it release H^+

- Attraction between nonbonding electrons on oxygen and the metal causes a shift of the electron density in water

    - This makes the O-H bond more polar and the water more acidic

    - Greater charge and smaller size make a cation more acidic

Effects of Cations and Anions

- An anion that is the conjugate base of a strong acid will not affect the pH

- An anion that is the conjugate base of a weak acid will increase the pH

- A cation that is the conjugate acid of a weak base will decrease the pH

- Cations of a strong Arrhenius base will not affect the pH

- Other metal ions will cause a decrease in pH

- When a solution contains both the conjugate base of a weak acid and the conjugate acid of a weak base, the effect on pH depends on the K_a and K_b values

CHAPTER 16 – SOLUBILITY EQUILIBRIUM

What is Solubility Equilibrium?

- If an "insoluble" or slightly soluble material is placed in water, an equilibrium forms between the undissolved solids and ionic species in solutions

    - Solids continue to dissolve, while ion-pairs continue to form solids

- Consider the reaction: $AgCl(s) \rightleftharpoons Ag^+(aq) + Cl^-(aq)$

    - $K = \dfrac{[Ag^+][Cl^-]}{[AgCl]}$

    - However, remember that since AgCl is a pure solid it isn't considered in K and thus the equation can be rewritten as: $K_{sp} = [Ag^+][Cl^-]$

Solubility Product (K$_{sp}$)

- General expression: $M_mX_n(s) \rightleftharpoons mM^{n+}(aq) + nX^{m-}(aq)$

    - Solubility product for the general expression: $K_{sp} = [M^{n+}]^m[X^{m-}]^n$

- Example of how to find solubility (s) from K$_{sp}$:

    - $AgCl(s) \rightleftharpoons Ag^+(aq) + Cl^-(aq)$

        - $K_{sp} = [Ag^+][Cl^-] = 1.6 \times 10^{-10}$

    - If s is the solubility of AgCl, then:

        - $[Ag^+] = s$ and $[Cl^-] = s$

        - $K_{sp} = (s)(s) = s^2 = 1.6 \times 10^{-10}$

        - $s = 1.3 \times 10^{-5}$ mol/L

- Another example:

    - $Ag_3PO_4(s) \rightleftharpoons 3Ag^+(aq) + PO_4^{3-}(aq)$

        - $K_{sp} = [Ag^+]^3[PO_4^{3-}] = 1.8 \times 10^{-18}$

    - If the solubility of Ag$_3$PO$_4$ is s mol/L, then:

        - $K_{sp} = (3s)^3(s) = 27s^4 = 1.8 \times 10^{-18}$

        - $s = 1.6 \times 10^{-5}$ mol/L

Factors Affecting Solubility

- Temperature
    - Generally, solubility increases with temperature
- Common ion effect
    - Common ions reduce solubility
    - Consider the following solubility equilibrium:
        - $AgCl(s) \rightleftharpoons Ag^+(aq) + Cl^-(aq)$; $K_{sp} = 1.6 \times 10^{-10}$
        - The solubility of AgCl is 1.3×10^{-5} mol/L at 25 °C.
        - If NaCl is added, equilibrium shifts left due to increase in $[Cl^-]$ and some AgCl will precipitate out
        - For example, if $[Cl^-] = 1.0 \times 10^{-2}$ M,
            - Solubility of AgCl = $(1.6 \times 10^{-10})/(1.0 \times 10^{-2}) = 1.6 \times 10^{-8}$ mol/L
- pH of solution
    - pH affects the solubility of ionic compounds in which the anions are conjugate bases of weak acids
    - Consider the following equilibrium:
        - $Ag_3PO_4(s) \rightleftharpoons 3Ag^+(aq) + PO_4^{3-}(aq)$;
        - If HNO_3 is added, the following reaction occurs:
            - $H_3O^+(aq) + PO_4^{3-}(aq) \rightleftharpoons HPO_4^{2-}(aq) + H_2O$
        - This reaction reduces PO_4^{3-} in solution, causing more solid Ag_3PO_4 to dissolve
- Formation of complex ion
    - Formation of complex ion increases solubility
    - Many transition metals ions have strong affinity for ligands to form complex ions
        - Ligands are molecules such as: H_2O, NH_3 and CO, or anions, such as F^-, CN^- and $S_2O_3^{2-}$

- o Complex ions are soluble

  - ▪ So the formation of complex ions increases solubility of slightly soluble ionic compounds

Predicting Formation of Precipitate

- $Q_{sp} = K_{sp}$

  - o Saturated solution, but no precipitate

- $Q_{sp} > K_{sp}$

  - o Saturated solution, with precipitate

- $Q_{sp} < K_{sp}$

  - o Unsaturated solution

- Q_{sp} is ion product expressed in the same way as K_{sp} for a particular system

- Example Question: 20.0 mL of 0.025 M $Pb(NO_3)_2$ is added to 30.0 mL of 0.10 M NaCl. Predict if precipitate of $PbCl_2$ will form. Given: K_{sp} for $PbCl_2$ = 1.6 x 10^{-5}

  - o $[Pb^{2+}]$ = (20.0 mL x 0.025 M)/(50.0 mL) = 0.010 M

  - o $[Cl^-]$ = (30.0 mL x 0.10 M)/(50.0 mL) = 0.060 M

  - o Q_{sp} = $[Pb^{2+}][Cl^-]^2$ = (0.010 M)(0.060 M)^2 = 3.6 x 10-5

  - o $Q_{sp} > K_{sp}$, so $PbCl_2$ will precipitate

CHAPTER 17 – ELECTROCHEMISTRY

What is Electrochemistry?

Electrochemistry is a branch of chemistry concerned with the study of the relationship between electron flow and redox reactions.

Review:

- Oxidation-reduction reactions (aka redox reactions) – reactions that involve a partial or complete transfer of electrons from one reactant to another

    - Oxidation = loss of electrons

    - Reduction = gain of electrons

        - Trick for remembering which is which - OIL RIG

        - **OIL - O**xidation **I**s **L**osing electrons

        - **RIG - R**eduction **I**s **G**aining electrons

    - Oxidation and reduction always occur simultaneously

Oxidation Number = valence e^- – (unbonded e^- + bonding e^-)

- Oxidizing agent – species that oxidizes another species

    - It is itself reduced – gains electrons

- Reducing agent – species that reduces another species

    - It is itself oxidized – loses electrons

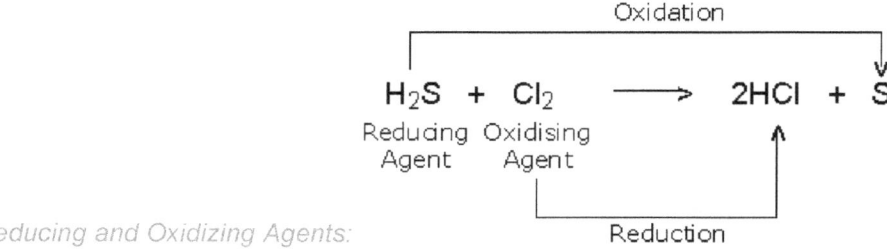

Reducing and Oxidizing Agents:

- Oxidation number (aka oxidation state) – actual charge an atom in a molecule would have if all the electrons it was sharing were transferred completely, not shared

Formal Charge vs. Oxidation Number

- Formal charges are used to examine resonance hybrid structures

    - Oxidation numbers are used to monitor redox reactions

- Formal Charge

    - Bonding electrons are assigned equally to the atoms

        - Each atom has half the electrons making up the bond

    - Formal Charge = valence e^- – (unbonded e^- + ½ bonding e^-)

- Oxidation Number

    - Bonding electrons are transferred completely to the more electronegative atom

Half-Reactions

- Redox reactions can be written in terms of two half-reactions

    - One involves the loss of electrons (oxidation)

    - The other involves the gain of electrons (reduction)

    - Example: $Fe^{2+} + Ce^{4+} \rightarrow Fe^{3+} + Ce^{3+}$

$$Fe^{2+} \rightarrow Fe^{3+} + e^- \quad \text{(oxidation half - reaction)}$$
$$Ce^{4+} + e^- \rightarrow Ce^{3+} \quad \text{(reduction half - reaction)}$$
$$\overline{Fe^{2+} + Ce^{4+} \rightarrow Fe^{3+} + Ce^{3+}}$$

- A balanced redox equation has to have charge balance

    - Number of electrons lost in the oxidation half-reaction must be equal to the number of electrons gained in the reduction half-reaction

Rules for Balancing Redox Reactions

Balancing Redox Equations with Ion-Electron Method or Half-Reaction Method

- Write two half-reactions and balance both for:

    - The number of the key atom (i.e. the atom changing oxidation numbers)

    - Change in oxidation number with electrons

- Add half-reactions so electrons cancel

- Balance charge with OH⁻ (if the reaction is occurring in a base) or H⁺ (if the reaction is occurring in an acid)

- Balance O atoms with H_2O

- Check that there is no net change in charge or number of atoms

Oxidation Number Method

- Determine oxidation number of atoms to see which ones are changing

- Put in coefficients so that no net change in oxidation number occurs

- Balance the remaining atoms that are not involved in change of oxidation number

- Example: Consider the reaction: $HNO_3 + H_2S \rightarrow NO + S + H_2O$

    - Oxidation numbers: N = 5, S = -2 → N = 2, S = 0

        - N goes from 5 → 2

            - Δ = −3 reduction

        - S goes from -2 → 0

            - Δ = +2 oxidation

    - Multiply N by 2 and S by 3

        - $2 HNO_3 + 3 H_2S \rightarrow 2 NO + 3 S + H_2O$

    - Balance O in H_2O

        - 6 (Ox) → 2(Ox) + 4 H_2O

    - Write the final reaction and make sure it is balanced (same number of atoms on left and right side)

        - $2 HNO_3 + 3 H_2S \rightarrow 2 NO + 3 S + 4 H_2O$

Voltaic (Galvanic) Cells

Voltaic cells are electrochemical cells in which a product-favored (spontaneous) redox reaction generates an electric current. The reaction produces an electron flow through an outside conductor (wire). Requirements for voltaic cells:

- Anode - an electrode (i.e. conductor such as metal strip or graphite) where oxidation occurs

- Cathode - an electrode where reduction occurs

- Salt bridge - tube of an electrolyte (sometimes in a gel) that is connected to the two half-cells of a voltaic cell

  - Salt bridge allows the flow of ions but prevents the mixing of the different solutions that would allow direct reaction of the cell reactants

  - Charge does not build up in half cells

    - Electrical neutrality must be maintained

Cell Diagrams

Cell diagrams are shorthand representation for an electrochemical cell.

- Anode is placed on the left side

- Cathode is placed on the right side

- Single vertical line represents a boundary between phases, such as between an electrode and a solution

- A double vertical line represents a salt bridge or porous barrier separating two half-cells

phase boundary   two phase boundaries   phase boundary

$$Zn(s) \mid Zn^{2+}(aq, + M) \parallel Cu^{2+}(aq, + M) \mid Cu(s)$$

anode   **salt bridge**   **cathode**

Electron Potential

- Electron flow in galvanic cell can do work/produce energy

- Electrical potential energy is measured in volts

  - 1 volt = (1 joule) / (1 coulomb)

    - Coulombs = amperes x seconds:

      - $C = A \times s$ or $A = C / s$

Standard Cell Voltages

- Cell voltages can be measured under standard conditions: 1 atm pressure, 25^0 C, and 1.0 M concentrations

  - Denoted as E^0_{cell}

- The standard cell potential is the sum of the standard potentials for the oxidative half-reaction and the reductive half-reaction

- If E^0_{cell} is positive, the net cell reaction is said to be product-favored (spontaneous)

- If E^0_{cell} is negative, the net cell reaction is said to be reactant-favored (nonspontaneous)

Standard Electrode Potentials

Standard electrode potentials are measured for half-reactions, relative to a standard hydrogen electrode potential (which has an assigned value of 0 volts).

- Each half reaction is written as a reduction

- Each half reaction could occur in either direction

- The more positive the standard electrode potential, the greater the tendency to undergo reduction

    o That means it is a good oxidizing agent

- The more negative the standard electrode potential, the greater the tendency to undergo oxidation

    o That means it is a good reducing agent

- If a half-reaction is written in the reverse direction, you must flip the sign of the corresponding standard electrode potential

- If a half-reaction is multiplied by a factor, the standard electrode potential is **not** multiplied by that factor

Cell Potential and Gibbs Free Energy

- Since a positive E^0_{cell} indicates a spontaneous reaction, you might imagine there is relationship between E^0_{cell} and free energy (ΔG^0)

- $\Delta G^0 = -nFE^0_{cell}$

    o n = # of moles of electrons transferred

    o F = Faraday constant = $9.65 \times 10^4 \dfrac{C}{mole \cdot e^-}$

    o ***Important***: A positive E^0_{cell} would result in a negative ΔG^0, and a negative E^0_{cell} would result in a positive ΔG^0

Electrolytic Cells

Electrolytic cells consist of an electrolyte, its container, and two electrodes, in which the electrochemical reaction between the electrodes and the electrolyte produces an electric current.

- Properties of electrolytic cell

    - Requires energy (in the form of an electric current)

    - No physical separation is needed for the two electrode reactions

    - Usually no salt bridge is required

    - Conducting medium is molten salt or aqueous solution

    - For electrolytic redox reaction:

        - E^0_{cell} is negative

        - ΔG^0 is positive

        - K_c is small (<1)

CHAPTER 18 - NUCLEAR CHEMISTRY

Radioactivity

Radioactivity is the emission of ionizing radiation or particles caused by the spontaneous disintegration of atomic nuclei.

- Types of radioactivity: alpha, beta, and gamma decay

  o Also positron emission

- Convention to be aware of:

Mass number $\searrow$

$^{A}_{Z}X \longleftarrow$ Chemical symbol

Atomic number $\nearrow$

Nuclear Equation

- Sum of the atomic numbers on both sides of the nuclear equation must be equal

- Sum of the mass numbers on both sides of a nuclear equation must be equal

Parent nuclide $\searrow$ Daughter nuclides $\swarrow \searrow$

Nuclear Equation Example: $^{238}_{92}U \longrightarrow {}^{234}_{90}Th + {}^{4}_{2}He$

Alpha Decay

- Alpha particles are nuclear decay particles

  o An unstable nucleus emits a small piece of itself

- Alpha particles consist of two protons and two neutrons

- Alpha particle symbol: α

  o An α particle is a helium nucleus

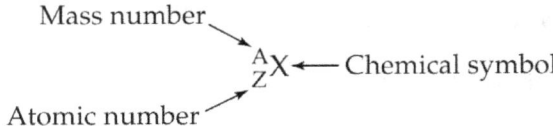

$${}^{4}_{2}He \ or \ {}^{4}_{2}\alpha$$

Alpha Particle:

- Alpha particles are ejected from the nucleus at a fairly low speed (approximately one-tenth the speed of light)

  o They are a minimal health risk to people unless ingested or inhaled

- Large mass nuclei tend to use alpha emission because it is a quick way for a large mass atom to lose a lot of nucleons (either a proton or neutron)

$$^{238}_{92}U \longrightarrow {}^{234}_{90}Th + {}^{4}_{2}He$$

Alpha Decay Equation Example:

Beta Decay

- Beta radiation symbol: β or $^{0}_{-1}e$

- Beta emission is a nuclear decay process that ejects a high speed **electron** from an unstable nucleus

- Electron is formed within the nucleus by the breakdown of a neutron into a proton and electron

    o The electron is ejected from the system

    o The proton that was formed remains behind in the nucleus

        ▪ Because of the addition of the proton, the atomic number of an element increases during beta emission

- Beta emission can be a significant health risk

$$^{14}_{6}C \longrightarrow {}^{14}_{7}N + {}^{0}_{-1}\beta$$

Beta Decay Equation Example:

Gamma Decay

- Gamma radiation symbol: γ

- Gamma emission occurs primarily after the emission of a decay particle

- Gamma is a form of high energy electromagnetic radiation

    o It is a significant health risk

- After a particle is ejected from a nucleus the system may have some slight excess of energy, or exist in a meta-stable state

- Gamma emission does not result in change of the isotope or the element

    o No mass and no charge change

Gamma Decay Equation Example: $$^{125}_{53}I^{*} \longrightarrow {}^{125}_{53}I + \gamma$$

        ▪ The asterisk is used to indicate that the element is in a high energy state

Positron Emission

- An unstable nucleus emits a positron

- A positron has the same mass as an electron but the charge is +1

Positron Emission Equation Example: $$^{15}_{8}O \longrightarrow ^{15}_{7}N + ^{0}_{+1}\beta$$

Half-life

Half-life is defined as the time for ½ of the parent nuclides to decay to daughter nuclides.

- All radioactive decay is first order

    o Rate = $-\frac{\Delta N}{\Delta t} = kN$

        - t – time

        - N - # of atoms

        - k = rate constant

    o $\ln(N_o/N) = k$

        - N_o - # of atoms at the starting time

- Half-life formula: $t_{\frac{1}{2}} = \frac{\ln(2)}{k} = \frac{0.693}{k}$

- Half-life is a constant

Carbon-14 Dating

- Carbon-14 dating can be used to date objects ranging from a few hundred years old to 50,000 years old

- Carbon-14 is produced in the atmosphere when neutrons from cosmic radiation react with nitrogen atoms

  - $^{14}_{7}N + ^{1}_{0}n \rightarrow ^{14}_{6}C + ^{1}_{1}H$

- Living things take in carbon dioxide and have the same ^{14}C to ^{12}C ratio as the atmosphere

  - However, when a plant or animal dies, it stops taking in carbon as food or air

  - Radioactive decay of carbon starts to change the ratio of $^{14}C/^{12}C$

  - By measuring how much the ratio is lowered, we can determine how much time has passed since the plant or animal lived

- Half-life of cabon-14 is 5,720 years

Fission and Fusion

- In fission, a large mass nucleus is split into two or more smaller mass nuclei

$$^{235}_{92}U + ^{1}_{0}n \rightarrow ^{139}_{56}Ba + ^{94}_{36}Kr + 3\ ^{1}_{0}n$$

Fission Equation Example:

- In fusion, small mass nuclei are combined to form a larger mass nucleus

$$^{2}_{1}H + ^{3}_{1}H \rightarrow ^{4}_{2}He + ^{1}_{0}n$$

Fusion Equation Example:

  - Fusion requires very high temperatures (in the millions of degrees) so that small nuclei can collide together at very high energies

CONCLUDING REMARKS

I hope this book has provided you tremendous value for your money and has helped you do better on your exams! If it has done both of these things, I have achieved my purpose in making this guide.

Furthermore, my goal is to create more books and guides that continue to deliver great value to readers like you for little monetary costs. Thank you again for purchasing this study guide and I wish you the best on your future endeavors!

- Dr. Holden Hemsworth

MORE BOOKS BY HOLDEN HEMSWORTH

DO YOU NEED HELP WITH OTHER CLASSES?

CHECK OUT OTHER BOOKS IN THE ACE! SERIES

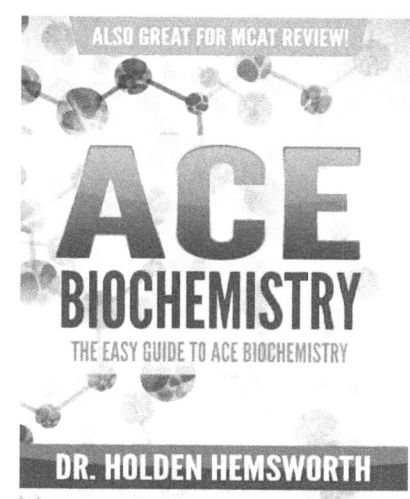 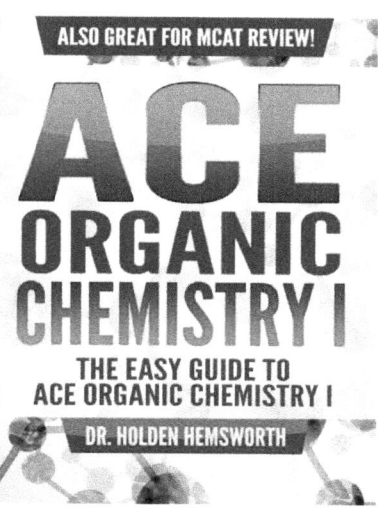 

ALL BOOKS ARE LISTED ON MY AMAZON AUTHOR PAGE

MORE BOOKS COMING SOON!